THE MIGHTY A: THE SHORT, HEROIC LIFE OF THE USS *ATLANTA* (CL-51)

THE MIGHTY A: THE SHORT, HEROIC LIFE OF THE USS *ATLANTA* (CL-51)

America's First Warship Commissioned After Pearl Harbor

DAVID F. WINKLER

Published in the United States of America and Great Britain in 2025 by
CASEMATE PUBLISHERS
1950 Lawrence Road, Havertown, PA 19083
and
47 Church Street, Barnsley, S70 2AS, UK

Hardback Edition: ISBN 978-1-63624-502-7
Digital Edition: ISBN 978-1-63624-503-4

A CIP record for this book is available from the British Library

Printed and bound in the United States of America by Integrated Books International

Typeset in India by Lapiz Digital Services, Chennai.

For a complete list of Casemate titles, please contact:

CASEMATE PUBLISHERS (US)
Telephone (610) 853-9131
Fax (610) 853-9146
Email: casemate@casematepublishers.com
www.casematepublishers.com

CASEMATE PUBLISHERS (UK)
Telephone (0)1226 734350
Email: casemate@casemateuk.com
www.casemateuk.com

Contents

Preface

On November 13, 1943, USS *Juneau* (CL-52) was ripped apart by a torpedo launched by Japanese submarine *I-26.* An enormous explosion followed by remnants of the light cruiser slipping below the surface in less than 30 seconds, stunned those on other warships in a flotilla that was steaming southward from Guadalcanal. Believing there could be no survivors and fearful of another torpedo attack, the flotilla pressed on. News of the loss of this ship was made all the worse after it was discovered there had been about 100 survivors left behind in shark-infested waters. Only 10 of a crew of 697 survived. Among those killed were the five Sullivan brothers from Waterloo, Iowa. To honor the sacrifice of those five brothers and their shipmates, the Navy named two destroyers for them with USS *The Sullivans* (DD-537) surviving as a museum ship at the Buffalo and Erie County Naval and Military Park in western New York, and USS *The Sullivans* (DDG-68) currently serving with the fleet. The Navy would also name another light cruiser (CL-119) for Alaska's capital. Launched late in World War II, this warship served during the Korean War. In addition, the Navy commissioned an amphibious warship (LPD-10) that carried on the name that served for three decades into the 21st century.

Built at the Federal Shipbuilding and Dry Dock Company of Kearny, New Jersey, *Juneau* was commissioned at the Brooklyn Navy Yard on Valentine's Day, 1942, seven weeks after her Kearny-built sister *Atlanta* (CL-51) joined the fleet at that location on Christmas Eve, 1941. In addition to sharing the same building yard and commissioning site, the two ships would also share the same November, Friday the 13th, demise with *Atlanta* being scuttled off Guadalcanal in waters that had become known as Ironbottom Sound.

The purpose of this narrative is to cover the story of that other Kearny-built light cruiser as a compendium to my recently published *Witness to Neptune's Inferno: The Pacific War Diary of Lieutenant Commander Lloyd M. Mustin*, USS *Atlanta* (CL 51). After reviewing the diary that had been loaned to me by Vice Adm. Henry C. Mustin in 2001 (and returned to his son Vice Adm. John B. Mustin in 2024) I saw the potential for a narrative that placed the observations of the future Vice Adm. Lloyd M. Mustin in context. Initially that included exploring and telling the story of the ship he served on—the USS *Atlanta* (CL-51). In addition to digging through the ship's files at the then Naval Historical Center (today Naval History and Heritage Command) and deck logs, and war diaries at the National Archives, I still had the opportunity to capture some of the recollections of some of the junior officers who had served in *Atlanta*. Traveling to Chicago I met retired journalist Ed Corboy who published a series of articles in the *Chicago Tribune* about the "*Mighty A*'s" exploits. Robert Graff, a retired media executive living in New Jersey, was also generous with his time. I also benefited from an extensive oral history that had been completed with Mustin thanks to the U.S. Naval Institute as well as oral histories that had been completed by the Atlanta History Center and Library of Congress Veterans History Project.

What emerged was a rather large manuscript that I felt was ready for prime time and I sought the help of James D. Hornfischer. Besides achieving acclaim for several popular World War II naval histories such as *The Last Stand of the Tin Can Sailors*; *Ship of Ghosts*; and *Neptune's Inferno*, Hornfischer served as a literary agent to help other authors find a publisher for their work. Reviewing the bulky manuscript that had the working title of *Mustin's Diary*, Hornfischer observed that the inclusion of ship history content did not work, as Lloyd Mustin's entries into a standard-issue government logbook focused mainly on the American conduct of the war. With the exception of praise for the commanding officer and harsh criticisms of the executive officer, there was not much on *Atlanta* (his oral history did go into more detail). Though rejection is never a happy outcome, Hornfischer pointed out that within my monograph there existed not one, but two books! Thanks to Casemate, *Mustin's Diary* was published in 2024 as *Witness to Neptune's Inferno* (a title that pays homage to the late Jim Hornfischer). To provide context for several entries referring to *Atlanta* in the diary, I discussed what the ship was doing so there is some content overlap with this previous book. My goal here was to be not excessive.

This book benefited from additional research, advice, and prior publication. For example, I presented content from chapter 1 originally as a paper titled "The Construction of USS *Atlanta* and the Navy Seizure of Federal Shipbuilding" at the 2012 Business History Conference held in Philadelphia. Incorporating feedback from that gathering, the paper would be published in 2014 in Volume 2 of *The Northern Mariner* journal of the Canadian Nautical Research Society. More recently, at the 2022 Western Naval History Association conference on board the aircraft carrier *Midway* in San Diego, I enjoyed a presentation by Lonnie Gill and his wargaming associates titled: "Wargaming Reenactment of the 13 November 1942 Naval Battle of Guadalcanal Based on U.S. Signal Traffic Using 1:2400 Miniatures." Subsequently, Lonnie was kind enough to lend his expertise to perform a sanity check on my "Friday the 13th" chapter.

I thank the team at Casemate for agreeing to publish this work. Of course a continuing thanks goes out to my wife Mary, and daughters Xepher and Carolyn for their support over the years.

David F. Winkler

Alexandria, VA, 2024

CHAPTER I

The Construction and Manning of the USS *Atlanta*

September 6, 1941, proved a memorable day at a shipyard in northern New Jersey as Margaret Mitchell arrived to christen America's newest cruiser, the *Atlanta*. The 40-year-old native of the ship's namesake city had achieved local notoriety in the state of Georgia as the author of nearly 130 articles for the *Atlanta Journal* magazine. However, her one book *Gone With The Wind*, published in 1936, became a national best seller and earned her a Pulitzer Prize in 1937. The 1939 movie adaptation of Mitchell's southern perspective of the Civil War earned eight Oscars including the coveted best motion picture award. Thus, four months earlier when Captain A. T. Bidwell, an assistant of Rear Adm. Chester W. Nimitz at the Bureau of Navigation, wrote to Mayor Roy LeCraw of Atlanta to seek a nominee to serve as the ship's sponsor, LeCraw quickly nominated Mitchell for the ceremonial honor. Nimitz, however, declined the mayor's suggestion that the ship be christened with a large bottle of Coca-Cola.[1]

Preparations for *Atlanta*'s slide down Building Way #8 at the Federal Shipbuilding and Dry Dock Company along the banks of the Hackensack River had begun at the time Mitchell was selected as sponsor, with the application of stearin lubricant on the run beneath the uncompleted cruiser's keel. A ⅜-inch film was applied on the run from amidships forward and the coat increased in thickness to half inch from amidships aft. A few days later, workers laid another half inch of "launching grease" atop of the stearin. On July 28, a diver greased and covered the underwater portion of the run leading out from the building way. Totaled together, 19,600 pounds of lubricant was applied to the run. On the day prior to the launch, a diver then waded into the Hackensack to remove the protective coverings. As he performed this chore, other yard workers began removing every other keel block from beneath the cruiser.

Beginning at 0400, prior to sunrise, on that Saturday morning workers moved grease covers and methodically pulled the remaining keel blocks and supports, called bents, away from the hull that sported a light gray coat above the waterline and a coat of black anti-fouling paint below. At 0900, a thorough inspection was made to assure watertight integrity. Yard workers closed off all sea openings and shut interior hatches from the second deck aft and below.

Arriving VIPs could see that only two cradles, fashioned out of 200 tons of yellow pine timber, kept the 542-foot-long ship upright in place. Decorations were modest. Yard workers draped red, white, and blue bunting over the bow and strung up signal flags from the forecastle to the top of the mast. Launching attendees joining Mitchell included the Governor of New Jersey, Mr. Charles Edison and his wife; Undersecretary of the Navy, James V. Forrestal; the Director of the Naval Research Laboratory, Rear Admiral Harold G. Bowen; and the Commander of the New York Navy Yard, Rear Admiral Edward J. Marquart. For Edison, the launching must have been gratifying as the ship's design was conceived and evolved during his tenure as secretary of the Navy. For Bowen, who had been sent to New Jersey a month earlier

Margaret Mitchell, the author of *Gone with the Wind*, was the sponsor of USS *Atlanta* (CL-51). (Archives Branch, Naval History and Heritage Command, Washington, DC; NH 47714)

to seize the shipyard from a corporate subsidiary of United States Steel, this was not where he wanted to be on this post-Labor Day Saturday.

The *New York Times* described the festive ceremony noting that Miss Mitchell wore "a black crepe dress with white lace collars and cuffs, black flat-heeled shoes, and black silk stockings" along with "a pink, off-the-face hat, and a diamond ring, a wristwatch, and pearl earrings." The author then grasped the bottle of Champagne "much as she would have held a baseball bat" and "cracked it squarely and neatly against the ship's knifelike prow, spattering most of the frothy contents well over the hard steel."

The bubbly splattered, triggers were released, and the cruiser then began a slow descent down the ways, picking up speed as she backed into the murky waters of the Hackensack. On board, 181 men, the majority being shipyard workers, rode with the ship as it reached a speed of 20 feet per second before the stern splashed into the river. Loud applause broke out and horns sounded. Afloat, *Atlanta* continued to back out until the stern rested some 750 feet from the end of the run. Four tugs saddled up against the cruiser, and workers struggled to drop the two cradles from beneath the hull. Once free of the cradles, the tugs pushed the ship bow in to berth two in the finishing basin to continue the installation of weapons and other equipment.[2]

Meanwhile, the launching party capitalized on another Atlanta connection as the local Coca-Cola bottling plant, located across from the yard, hosted a lunchtime reception where Lt. Cdr. Norman W. Sears, the acting commanding officer of the pre-commissioning unit and future first lieutenant, accepted a silver punch bowl presented on behalf of the citizens of Atlanta.[3]

Atlanta (CL-51) was the third American warship to inherit the name of the capital of Georgia. The first *Atlanta* began service as the British iron-hulled steamer *Fingal*. Metallic sides, combined with an ability to make 13 knots, made the 185-foot-long, 800-ton vessel an attractive buy for southern purchase agents shopping in England during the summer of 1861. Loaded with gunpowder, armaments, and other supplies, *Fingal* slipped through the Union blockade and moored at Savannah on November 12, 1861. The cross-Atlantic journey would be the steamer's last, as Union Navy warships blocked the Savannah River egress. Recognizing her seagoing days were past, the South converted the steamer into an ironclad, razing the hull to nearly the waterline from which rose a sloped casemate that would house four heavy Brooke rifles.

Christened the *Atlanta* in July 1862, the now 1,000-ton vessel underwent engine trials and awaited further orders. With Union troops ashore, underwater obstructions, and Union Navy monitors blocking *Atlanta*'s direct access to the sea, the ironclad maneuvered through a series of backwater canals and creeks to emerge into Wassaw Sound on June 17, 1863. The Union Navy sent in the ironclad monitors *Weehawken* and *Nahant* to confront the rebel ship. Unfortunately for the southern seamen, the Achilles heel of their warship was her deepened draft. When she maneuvered to confront the oncoming enemy, she grounded not once, but twice, with the second grounding causing her to heel over, making her guns virtually useless. Firing 15- and 11-inch Dahlgren guns, the two Union monitors conducted target practice on the hapless southern ironclad with telling effect. Within minutes, she struck her flag.

Salvaged, *Atlanta* underwent repair and rearmament at the Philadelphia Navy Yard and was placed into commission in December 1863 as a U.S. Navy warship, assigned to the North Atlantic Squadron in the Hampton Roads area. At the conclusion of the war, the ship was laid up, later sold, and broken up for scrap in 1869.[4]

Congress authorized the construction of the second *Atlanta* and three additional ships in 1883 to be the vanguard of the new "steel navy." Built by John Roach and Sons at Chester, Pennsylvania, *Atlanta* was designed as a protected cruiser. The Navy commissioned the modern warship on July 19, 1886, at the New York Navy Yard. Despite her modernity, *Atlanta* still featured masts and yardarms for sailing. So rigged, she proceeded to represent the United States on several overseas deployments to Europe and South America. Placed out of commission in 1895, *Atlanta* missed the Spanish-American war but would reenter service, minus the yardarms, in 1900. Following extensive service in the Caribbean, the obsolescent cruiser spent her last seven years as a barracks ship before being struck and sold in 1912.

Nearly three decades would pass before another ship named for the capital of the Peach State would be commissioned. A combination of

The two *Atlanta* predecessors: CSS *Atlanta* would be captured and serve in the U.S. Navy, and USS *Atlanta* would be one of the first vessels of the new "steel" navy. (Archives Branch, Naval History and Heritage Command, Washington, DC; NH 57819; NH 91653)

international and domestic events created the requirement for the United States Navy to build a new type of light cruiser.

Through the 1930s, while the world's economies were mired in depression, overseas conflicts concerned the administration of President Franklin D. Roosevelt. Japan and China were fighting on the Asian mainland. In Europe, Adolf Hitler gained control in Germany and Joseph Stalin ruthlessly liquidated perceived enemies within Russia. Spain was embroiled in a brutal civil war. Benito Mussolini's forces seized Ethiopia, ignoring the threat of sanctions imposed by the League of Nations.

Naval treaties negotiated between the great naval powers, first in Washington in 1922, and then in London in 1930, had limited the number of and the tonnage of the ships the United States could add to its Navy. Even then, tight budgets kept the American order of battle below the treaties' limits. Other naval powers, especially Japan, kept their navies at the prescribed limits. Seeking to further increase their naval power, the Japanese announced in August 1934 that beginning in 1936, they would no longer be bound by the limitations set by the treaty. That year the Americans, French, and British met in London to give naval arms control one more go over. Minor agreements were reached that would influence some of the U.S. Navy's building programs.

Due to the extreme cost-cutting measures made by the Herbert Hoover administration to restore the economy, the U.S. Navy had been grossly underfunded. Many of the Navy's ships were tied up pierside for lack of fuel. Budget constraints were such that the Navy Department instructed the Naval Academy to send the bottom half of the Class of 1933 home after May's graduation ceremonies with no commissions. Eventually, most of the midshipmen of the Class of '33 did receive their commissions, a day ahead of the Class of '34![5]

In President Franklin D. Roosevelt, the Navy had a friend. The former assistant secretary of the Navy had an appreciation of sea power's importance for America's economic health and national security. Besides getting its midshipmen commissioned, the Navy began to benefit from the Roosevelt administration in other ways. In 1933, Roosevelt arranged for public works funds appropriated under the National Industrial Recovery Act to be used for naval shipbuilding. Hence, the United States built the carriers *Enterprise* and *Yorktown* as public works projects. Yet, the use of public works funds for naval shipbuilding didn't sit well with many in Congress. That branch of government, however, also provided Roosevelt and the Navy a powerful ally—Carl Vinson, chairman of the Naval Affairs Committee.

Under the Georgia representative's leadership, the Vinson-Trammell Act of 1934 passed the House and Senate, and further enhanced the naval recovery. Under this act, the Navy constructed the carrier *Wasp*, and numerous cruisers, including the *Phoenix*, a *Brooklyn*-class light cruiser that eventually ended its career in the depths of the South Atlantic during the 1982 Falklands War as the Argentine Navy's *General Belgrano*. Perhaps more significantly, the funds enabled the Navy to construct capable submarines and build new classes of destroyers to replace the four-stacker flush-deckers of World War I vintage.

The December 1937 Japanese attack on the gunboat *Panay*, three Standard Oil tankers, and British gunboats on the Yangtze served to further concerns about the potential for war in the Far East. In Congress, Chairman Vinson again sought funds for additional shipbuilding. The German absorption of Austria in March 1938 served to help Vinson's cause.[6]

Hence the Vinson Act of 1938 authorized expanding the Navy even further. Under this Act, for example, the *Iowa*-class battleships that would serve into the 1990s and the aircraft carrier *Hornet* of Doolittle Raid fame, were authorized. However, when Congress appropriated funds on April 26, 1938, for construction of *Atlanta* and a sister ship, to be the *Juneau*, the Navy cited the earlier Vinson-Trammell Act of 1934 as authorization since the 1934 legislation provided for the replacement of obsolescent cruisers. Two months later, President Roosevelt signed into law an act making up for ship construction oversights made in the earlier Naval Appropriations Act. Under the 2nd Deficiency Act of June 25, 1938, two additional cruisers of the class were funded. They would be named *San Juan* and *San Diego*.[7]

The ship design that eventually evolved to become *Atlanta* had its roots in the naval arms control milieu of the mid-1930s. Great Britain and the United States had long haggled over the size of ships that should be permitted under the treaty regime as the United States favored larger, more powerful ships to cover the expanse of the Pacific Ocean. When the powerful, 10,000-ton *Brooklyn*-class appeared in 1933, the British warned that such ships could spawn another arms race and recommended that a 7,000-ton limit be set for future construction. In informal talks with the British in 1934, the Americans acceded to the British Admiralty's wishes on the cruisers, although they maintained their position on having battleships of 35,000 tons.

Talks in London in 1936 produced an informal understanding that future cruisers would displace no more than 8,000 tons and carry armament no heavier than 6-inch guns. Navy planners understood they could proceed with building a new class of cruisers as the 1934 Vinson-Trammell Act authorized the Navy to build ships to replace obsolete vessels. The *Omaha*-class, with the lead ship launched in 1919, was approaching block obsolescence.[8]

With the London Naval Conference having concluded in May 1936, the Navy's General Board sent out a survey to the principal flag officers regarding the missions and characteristics of a new medium cruiser. Responses back validated a 1933 General Board policy document that argued that medium cruisers should be designed with the forethought of conducting combat with a secondary priority assigned to performing a long-range sea control mission.

Given the doctrine at the time centered on the battleship line as the key projector of sea power, the combat role of the medium cruiser was interpreted as a protection mission, fending off attacks by destroyers, submarines, and aircraft. Most of the Navy's senior leaders agreed with the need for the new ship to perform the mission although they disagreed on the size and aspects of the ship's armament. While some called for a lighter version of the 10,000-ton cruisers in production to have a similar role with lighter armaments, others thought the new ships should be specialized to work as destroyer flotilla leaders.[9]

In December 1936, the Chief of Naval Operations, Adm. William H. Standley, decided that 10 cruisers of 5,000–7,000 tons should be built, and the General Board asked the Preliminary Design shop to draft some proposals.

The designers made speed a priority, and depending on the displacement, they speculated that the new ship could attain speeds between 34 and 36 knots. It was assumed that the ship would be armed with 6-inch 47-caliber guns,[10] however, the designers put forward sketches that also featured 5-inch 38 caliber mounts. Of the nine sketches put forward, one came close to the final design of the *Atlanta*.

A June 9 hearing of the General Board saw Adm. Thomas Hart argue for a 7,000-ton ship that could serve as a destroyer leader, but also have enough firepower to stand up against enemy cruisers. A consideration regarding armament was that the 6-inch 47-caliber guns were still in development and that a mixed battery of 5- and 6-inch guns would be inefficient on a smaller hull. Subsequently, at a 14 June hearing, Rear Adm. William S. Pye suggested an all 5-inch 38-caliber gunship. Others argued to retain the 6-inch mounts. The preliminary design people went back to work. In September, Rear Adm. W. R. Furlong at the Bureau of Ordnance concluded that a 6-inch 47 mount would not be available for some time.

Given that limitation, a General Board memo dated December 1937 recommended going ahead with the all 5-inch cruisers but to limit the class to four ships, with the thought that development of the 6-inch mount could facilitate a follow-on ship class. Standley's replacement as CNO, Adm. William D. Leahy, signed a letter specifying the design requirements on March 2, 1938.[11]

Atlanta's preliminary design was completed in July 1938. President Roosevelt expressed interest and reviewed a set of blueprints a month later. The president must have been impressed with the handsome design and the ship's fine lines. Not everyone was satisfied, however. The initial plans lacked sonar and catapults for floatplanes. The design would be modified to incorporate a sonar set but no space could be allocated for floatplane catapults.

To attain the desired high speeds, the Navy contracted with Babcock and Wilcox, which had a positive track record providing high-pressure boilers for new classes of destroyers coming down the ways. The goal was for the ship, at 7,400 tons, to reach 32.5 knots.

The designers reintroduced an alternating engine room/fire room setup that had been popular in previous decades. One 250kw diesel engine was placed aft to provide emergency power.[12]

Atlanta's design created a vessel unmatched on the planet in its ability to fire an antiaircraft broadside. With its heavy antiaircraft armament and electronic capabilities, the designers estimated berthing would have to accommodate 26 officers and 523 enlisted, as well as an additional 58 men if the ship were to serve as a flagship for a destroyer flotilla. When completed, *Atlanta*, would hold 35 officers and 638 enlisted. To take on the additional bodies, the mess deck was reduced, and the crew had to eat in shifts.[13]

The Navy sought bids for hulls 51 and 52 (*Atlanta* and *Juneau*) from Bethlehem Steel Company (Shipbuilding Division) at Quincy, Massachusetts, and from Federal Shipbuilding and Dry Dock Company at Kearny, New Jersey. When the bidding deadline passed on February 15, 1939, the Navy only received a bid from Federal proposing a cost of $15,950,000 for one ship or $14,200,000 if both ships were awarded to the New Jersey shipyard. Bethlehem Steel did bid for the next two ships of the class. The Navy awarded the contracts for *Atlanta* and *Juneau* to Federal Shipbuilding for $12,226,000 per ship on April 10, 1939, and the contract was signed 15 days later with a delivery date set for 1942. While Federal Shipbuilding and Dry Dock Company would build *Atlanta* and *Juneau*, Bethlehem Steel earned the contract for *San Juan* and *San Diego*.[14]

Located on the west bank where the lower Hackensack River empties into Newark Bay in northern New Jersey, the Federal Shipbuilding and Dry Dock Company's shipyard was one of America's most modern. Constructed during World War I as a United States Steel subsidiary, the yard provided small-scale merchant ships for the Allied cause. With the conclusion of the war and cancelation of contracts, the shipyard let go of thousands of workers. Following the onset of the Depression, what little work there was vaporized, and the yard could only sustain a workforce of 600.

Although having a Kearny postal address, the shipyard was actually located closer to the blue-collar neighborhoods of Jersey City across the river. Workers commuted to the shipyard from homes located throughout New Jersey, and in some cases, New York.

The election of Franklin D. Roosevelt certainly benefited the shipyard. In 1933, the shipyard received contracts to build *Mahan*-class destroyers using funds made available through President Roosevelt's National Industrial Recovery Act. A year later, the Navy used funds appropriated through the Vinson-Trammell Act of 1934 to contract for the heavy destroyer *Somers*. By 1936, some 3,000 workers fabricated new merchant ships and the yard earned additional contracts to build Navy destroyers. With its expanding workforce, Federal Shipbuilding earned praise for building ships quickly, and efficiently.[15]

While appearing to be a stellar operation on the surface, Federal Shipbuilding management's relationship with its workforce made the industrial facility a prime candidate for union organizers. The emergence of unionism at the shipyard during the 1930s would eventually affect the timetable of *Atlanta*'s commissioning in 1941.

Going back several decades, United States Steel had fervently opposed the unionization of its workers. However, the Industrial Union of Marine and Shipbuilding Workers of America (IUMSWA) earned the allegiance of the majority of the workforce in 1937, and established Local 16 to confront management with a host of demands, the ultimate being the creation of a closed-shop situation where employment at Federal Shipbuilding was contingent on joining the union.

Local 16 could have been more aggressive in pursuing its objectives, however, the situation overseas mitigated against calling for strikes. During the previous September, Germany had quickly invaded and conquered much of Poland, leaving eastern portions to the Soviet Army. Preparations went ahead to build a new cruiser within Building Way #8. On April 22, 1940, the keel—the backbone of the ship—was laid and soon steel ribs began appearing fore and aft.[16] Nearly three weeks later, 75 Wehrmacht divisions, with the close support of the Luftwaffe, invaded the Netherlands, Luxembourg, and Belgium. Following the evacuation of the British Army and other Allied soldiers at the French port of Dunkirk, the French government agreed to terms on June 22, leaving Britain and its Empire alone to stand up against Hitler.[17]

In the United States, one reaction to the fall of France, was to further increase defense spending. In mid-July the president signed "The Two-Ocean Navy Bill," opening the spigot to construct new shipyards and ancillary industrial facilities, as well as the expansion of current yards to support the unprecedented building program. For the Federal Shipbuilding and Dry Dock Company, contracts were signed with the Navy to build additional destroyers, cruisers, troop transports, and cargo ships. Meanwhile, construction accelerated on the destroyers and cruisers currently under contract. Work on *Atlanta* moved ahead of schedule.[18]

As the war in Europe expanded in 1941 with the Axis invasion of the Balkans and then the Soviet Union, the infusion of federal funds to the shipyards trickled down into the pockets of those working on the building ways, meeting union demands for fair compensation. An Atlantic Coast Standards Agreement which boosted wages by 12 percent and affected shipyards along the East Coast, had the approval of both the company and the union, and had the blessings of Navy Secretary Frank Knox; Emory S. Land of the Maritime Commission; and the Office of Production Management.

However, Local 16's demand for a closed shop and the reclassification of some of the jobs within the shipyard remained contentious issues. To resolve labor disputes potentially disruptive to the nation's mobilization for war, the Roosevelt administration established a Defense Mediation Board. After hearing an appeal from Local 16, the Board recommended the adoption of a modified union shop with grievance boards to be convened to address the reclassification issues. While short of a ruling making the Kearny shipyard a closed shop, the modified union shop arrangement required the company to fire any union members who allowed their dues to lapse. The ruling, if implemented, would virtually guarantee Local 16's ability to retain its current strength as a sought to recruit the non-members. If the Defense Mediation Board's objective was to maintain labor peace at Kearny, its ruling had the opposite effect. The management at Federal Shipbuilding and Dry Dock Company refused to act on the Board's recommendation. On August 7, the workers at the shipyard went on strike.[19]

Margaret Mitchell was supposed to be in New Jersey for the launching of *Atlanta* on August 9. The union offered to provide workers to allow the festive event to move forward. The company spurned the offer. *Atlanta* would remain high and dry indefinitely.[20]

Further talks between the government, management, and labor made no headway. On August 23, President Roosevelt signed the executive order authorizing the Navy's seizure of the shipyard, blaming the company's management for its failure to abide by the Mediation Board's recommendation.[21] Both the union and the company pledged to cooperate with the government, consequently, there was no need to send in military forces to seize the yard. Instead, Rear Adm. Harold G. Bowen left Washington the next day to take charge of the yard.[22]

Bowen and his staff arrived in Kearny on Sunday, August 24. By Tuesday, the yard was back in full production under Navy management.

These two images illustrate the progress made on the building way at Federal Shipbuilding and Dry Dock Company during the summer of 1940. (Archives Branch, Naval History and Heritage Command, Washington, DC; 19-N-27297; 19-N-27295)

The launching of *Atlanta* was set for September 6. Although the workers went back to work, the strike had impeded the progress of the ship's construction in more ways than just time lost. Arriving the day after the Navy assumed control of the shipyard, the prospective Assistant Gunnery Officer, Lt. Lloyd M. Mustin, gazed at *Atlanta*'s gun mounts, which had been delivered to the yard and had been left exposed to the hot summer humidity and thunderstorms. "There was a lot of rust in critical areas, things of a sort the shipyard didn't even know how to cure. I could see we were eventually going to have to do it ourselves and eventually we did."[23]

Mustin would join Sears as two of the early arrivals of the pre-commissioning crew. Mustin, a graduate of the Naval Academy Class of 1932, came from a family that had served in the Navy for generations and from a well-known father, Henry C. Mustin, who gained fame as the first to be catapulted off an underway warship back in 1915. A strong advocate for naval aviation, his naval career was cut short due to an aortic aneurysm—a ballooned artery that rubbed against his ribs. On August 23, 1923, the condition claimed his life.[24] The Navy not only honored the late Navy captain by naming an airfield for him in Philadelphia, but also by naming a destroyer for him. Meanwhile, Sears had graduated from the Naval Academy with the Class of 1925, seven years ahead of Mustin. A son of a fisherman, from September 1937 to July 1939 he achieved high-level visibility with his command of the presidential yacht *Sequoia*.[25] Both officers, veterans of the Asiatic Fleet, were married and had young children.

Other future members of the wardroom began to arrive at the two-story building set aside for the pre-commissioning crew near the cruiser being fitted out. Campbell Dallas Emory, the "XO," was born in 1899 to George Meade Emory and Josephine DeWolfe Poe, and raised in Seattle. As hinted by his father's surname, Emory boasted a military pedigree comparable to Mustin's. A descendant of Benjamin Franklin, Emory shared the same name as his grandfather who served as Aide-de-Camp for Maj. Gen. George Meade during the Civil War. His great-grandfather was noted topographer Maj. Gen. William H. Emory. Called "Dal" by his fellow Mids in the Class of '21, his *Lucky Bag* yearbook entry described him as the "prettiest little pink-cheeked lad that ever carried a cup of Piper-Heidsieck." Yet his photograph portrayed the image of a serious young man, and that demeanor would continue as he matured. "He was a very intense fellow," recalled Bob Graff. As was the case with the skipper, Emory found himself on a battleship for his first tour and also would see duty off Nicaragua and in the Asiatic. He returned home from 1933–35 as a Naval Reserve Officers Training Corps (NROTC) instructor at the University of Washington. There he married Elisabeth Morrison. The newlyweds traveled cross-country to Bethlehem Shipbuilding at Quincy, Massachusetts, where Emory fit out and commissioned *Phelps* as the destroyer's first XO. After a shore tour in Washington, DC he assumed command of the destroyer *Mayo* and then received orders to be the second in command on *Atlanta*.[26]

The doctor was also a commander. T. F. Cooper was the senior officer in a medical department. Other officers in his department included a medical officer Lt. Cdr. Carl C. Carver and a dentist, Lt. Robert F. Erdman.

The supply officer or "pork chop" was Lt. D. C. T. Grubbs, Jr. Lt. (jg) James D. Koiner assisted Grubbs and served as the disbursing officer.

With the exception of the medical and supply officers, the department heads were also all of that school on the Severn. As with the aforementioned First Lieutenant and Damage Control Officer Sears, the Navigator, Lt. Cdr. James S. Smith, Jr. was a fellow New Englander out of the Class of '25. He also knew Mustin from the time they served together in the Asiatic Fleet flagship *Augusta* in the mid-1930s, then commanded by Capt. Chester W. Nimitz.

The Gunnery Officer, Lt. Cdr. W. Richardson David "Bill" Nickelson and the Engineering Officer, Lt. Cdr. Arthur E. Loeser graduated with the Class of '27. The former hailed from Arkansas and the latter was a Jersey boy. While Nickelson's first tour was on a battlewagon, Loeser found himself fitting out the carrier *Saratoga*. Both men spent time at Pensacola in 1929 to obtain flight training, however, aviation was in neither man's future career plans. Instead both men pulled sea tours in a variety of warships, becoming experts in gunnery and engineering. For example, Nickelson served on five previous ships as either a gunnery or torpedo officer.[27]

Because the majority of the men on the ship were assigned to either the gunnery or engineering departments, both department heads had senior lieutenants as assistants. In the case of gunnery, it was Mustin and John T. Wulff held down the assistant engineer job. Originally from Illinois, "Jack" Wulff graduated in the Class of '31, a year ahead of Mustin. Assignment to the engineering departments on two aircraft carriers followed studies in marine engineering at the Naval Postgraduate School. The Massachusetts Institute of Technology prepared him well for the *Atlanta* assignment and he was one of the first to arrive in Kearny in June 1941 to observe the installation of the cruiser's propulsion plants.

Philip T. Smith, Jr., from Connecticut graduated with the Class of 1930 and served as the communications officer and held the important collateral duty as senior watch officer—responsible for setting up watch rotations at sea and in port. After his initial battleship tour following graduation from the Naval Academy, Smith had tours on five different ships before reporting in early June 1941 to Federal Shipbuilding.

Next in the hierarchy were the three mid-grade lieutenants. Girard L. "Pat" McEntee, graduated from Annapolis in 1935. James C. Shaw and Van Ostrand Perkins left three years later. All three led sailors in the gunnery and deck departments.

Shaw and Perkins were best friends, and both were happy newlyweds. Shaw of Minnesota, and Perkins of Connecticut, had served their initial tours on destroyers in the Asiatic Fleet and shared liberty together in various ports throughout the Orient. The bond between the two men extended with their young brides. Both women had sent their engagement announcements out on the same day, wed their husbands in Manila, considered themselves "Army Brats" and all four pairs of parents were married in 1912. Elizabeth "Betts" Perkins and Jane Shaw would form the core of a tight *Atlanta* wives club that would become a critical support group for the warship.

Reporting to the shipyard shortly after the launching, the two junior officers would be together for only a short period as, in October, Shaw was dispatched for six weeks of gunnery school at the Washington Naval Gun Factory located on the northern banks of the Anacostia River in the nation's capital. Perkins, however, stayed with the ship and trained to be the assistant damage control officer and repair officer. A firefighting training facility was set up on the property. There, Perkins led and trained pairs of hose teams, with one team battling the flames of a burning oil pit while the second team protected the first by spraying a fog of water over them. Engulfed in thick black smoke, the men trained using oxygen breathing apparatus equipped with canisters that produced oxygen through a chemical reaction.

When not fighting blazes, Perkins studied the ship's blueprints, especially the plumbing systems. He also learned about the capabilities of the equipment being installed in what would be his machine shops. If something broke, *Atlanta* needed to have the capability to fabricate replacement parts.[28]

At the bottom of the wardroom hierarchy were the junior officers or "JOs." In contrast to the middle and upper line officer leadership, most of these "jgs" and ensigns did not call the Naval Academy their alma mater but had attended colleges in the civilian sector and joined the Navy to receive commissions through the V-7 program that was established on June 25, 1940.[29] Nearly all attended schools with outstanding reputations: Boston College, Georgetown, George Washington University, Harvard, and Yale.

Two of those officers arrived on the banks of the Hackensack River on September 22 were graduates of the Midshipman School at Northwestern University. Ensigns Edward D. Corboy and G. (George) Bowdain Craighill, Jr. having earned their single gold stripes on March 14 and June 12, 1941, respectively, stayed on at the Evanston university to serve as navigation instructors to following classes. Corboy, a Chicago native, had completed his junior year of college in the summer of 1940 when he learned of the Navy's new commissioning program. Four years earlier, when he was a high school junior, Corboy had won an appointment to the Naval Academy, but "at that time my feet were comfortably, but non-militarily flat."[30]

Born in 1914, Craighill was older and more educated than Corboy. A Washingtonian, Craighill attended St. Albans Prep School and then left for college at the University of the South at Sewanee, Tennessee. He returned to the nation's capital to attend the George Washington University Law School. He then practiced law for a year with his father's law firm before deciding to obtain a commission in the Naval Reserve.[31]

Corboy recalled in his case, after taking the oath to enter the service, he showed up at the Chicago Naval Armory with a suitcase and shortly found himself on an overnight train to New York. Arriving in Gotham, he and some of his train bunkmates caught a cab to a pier at 132nd Street on the Hudson River which permanently berthed the former battleship *Illinois*, which had been converted to a barracks ship and renamed the *Prairie State*. Along with hundreds of other officer candidates, Corboy climbed up an accommodation ladder up the port side, crossed over to the starboard side of the old battlewagon, and climbed down another ladder to where motor launches lined up to take the young men to three battleships anchored out in the river. The Chicagoan found himself in a boat with 65 other officer candidates heading to *Wyoming*. Exchanging his civilian threads for apprentice seaman's uniform, Corboy began a one-month on-the-job training experience as *Wyoming*, along with the battleships *Texas* and *Arkansas* departed for a one-month Caribbean cruise. Returning to New York in late September 1940, Corboy recalled that approximately 80 percent of the candidates were offered an opportunity to attend a three-month Midshipman School while the remaining 20 percent were declined for physical or aptitude reasons.

Corboy returned to Chicago to complete his senior year. After finishing his fall semester, however, he received a call to report to Abbott Hall on the campus of Northwestern University. There, he received a concentrated dose of courses focusing on engineering, gunnery, navigation, communications, and deck seamanship. Corboy held the instructors, retired naval officers brought back on active duty, in high esteem. Having excelled in navigation, Corboy was invited to stay on at Abbott Hall after he received his commission in March to teach follow-on classes while he awaited his permanent orders. One of his star pupils was Craighill who joined him as a navigation instructor upon his commissioning in June. Both received their orders to the *Atlanta* on September 6.

After another train ride to the East Coast, the two ensigns arrived at the sprawling northern New Jersey industrial complex, 10 days after the christening. Besides the *Atlanta*, they noted a half dozen destroyers, *Atlanta*'s sister *Juneau* and several cargo ships in various stages of fabrication up on the building ways. Catching their first glance at their future home, Craighill and Corboy noted that electrical cables, acetylene hoses, and pneumatic gear snaked over *Atlanta*'s decks as riveters and welders continued work on the ship's interior and exterior. Corboy was impressed: "The sunlight striking her mighty turrets was a stirring sight ... it was inspiring to know that the ship bearing those turrets was to be the fastest and most graceful cruiser in any navy."

Reporting to a two-story building that housed the offices of the pre-commissioning crew, they met Commander Emory who now served as the senior officer present. Emory proceeded to introduce Craighill and Corboy to the other officers.[32] They first met Sears, who was busy double-checking the hatch fittings, fire pumps, and damage control communication circuits. Lieutenant Commander Loeser and his

Atlanta being fitted out on October 1, 1941, at Federal Shipbuilding and Dry Dock Company. (Archives Branch, Naval History and Heritage Command, Washington, DC; 19-N-27302)

Assistant Lieutenant Wulff were busy readying what would become their engineering plants. They met the Communications Officer, Lt. Paul T. Smith, who impressed Corboy as "a wise man with an Irish twinkle in his eye." The two newcomers then exchanged greetings with two officers facing enormous challenges: the Supply Officer, Lt. D. C. T. Grubbs and his assistant Ens. Jim Koiner. The two "Pork Chops" had to figure out what spare parts should be on hand for a ship that was first of her class and purchase provisions to eventually sustain a crew of 700. The XO introduced Craighill and Corboy to the Gunnery Officer, Lieutenant Commander Nickelson, who would be Corboy's boss and his assistant, Lieutenant Mustin. Nickelson welcomed the two to his team. Lieutenant Mustin gave each a hearty handshake and took the two along for a mini-tour of the guns and fire-control equipment, showing off the armaments with the same pride "a mother would introduce her children." Along the way they met three other officers, Lieutenant (jg) McEntee, and the aforementioned Jim Shaw, and Van Ostrand Perkins.

For the next two weeks, the two ensigns participated in daily inspections and tracked the ship's progress. They then found themselves assigned to the Communications Station in New York, reviewing and updating *Atlanta*'s communication publications. Craighill continued to perform this tedious chore as Corboy joined Shaw in Washington at the gunnery school.[33]

During Shaw's, Craighill's, and Corboy's time away, the wardroom of the future ship continued to grow. One of the new arrivals was Robert Graff. Another V-7 program graduate, Graff had just started his college education at Harvard when war broke out in Europe in September 1939. Having experience in sailing small boats while growing up in Scarsdale, New York, Graff tried to enlist in the Navy, but with enlistment quotas full, he was turned away. With the announcement of the V-7 program in 1940, Graff signed up and took his 30-day cruise on the battleship *New York* to the Panama Canal and back. After returning to Harvard to complete his degree in economics, Graff attended the three-month Midshipman School on the *Prairie State* along the Hudson; received his Naval Reserve commission; received orders to be assigned to *Atlanta*'s pre-comm crew; and reported for duty in Kearny.

Driving from his parents' home in Scarsdale for what became a daily commute, Graff parked his car and found the "dingy dusty office full of officers" where he found Commander Emory. "He quizzed me about my background and asked if I would be interested in communications," Graff recalled. Accepting the offer, Graff received a clearance badge that allowed him to walk onto the construction site. "Go over there and find your way to the signal bridge and take a look at it ... that will be your domain," Emory told the new ensign. After visiting the topside area near the bridge, Graff returned, and the XO introduced him to Perkins who would serve as his mentor.[34]

Along with the officers, enlisted sailors showed up in Kearny with orders in hand for assignment to the cruiser under construction. Some had transferred from other Navy ships or shore stations while most had been in the Navy for less than a year, having attended Navy basic training and in many cases, further training at Navy technical schools.

Although the Selective Service Act was passed in 1940, authorizing the draft, many of *Atlanta*'s young "boots" had enlisted in the naval service for a variety of reasons ranging from patriotism to the old adage of wanting to "see the world." At the time, the Navy operated basic training facilities at San Diego, California; Great Lakes north of Chicago, Illinois; Newport, Rhode Island; and Norfolk, Virginia. Due to their closer proximity, many of *Atlanta*'s bluejackets graduated from the two latter installations.

To help connect the ship to its namesake city, the Navy recruited sailors from Georgia with an incentive that they would serve on the ship to be named for the state capital. Retiring in 1972 as a Master Chief Boatswain's Mate, William T. Garner recalled: "I enlisted in the City of Atlanta in November 1941 along with several other young men ... We knew the *Atlanta* was going to be a very special type of fighting ship because the Navy had Gene Tunney, the ex-heavyweight champ down to Atlanta, Georgia, to have lunch with us and to swear us in."[35] Clifford Dunaway of Dallas, Georgia, saw an the article in the newspaper about the *Atlanta* and saw his opportunity to achieve his youthful goal of joining one of the sea services. As a young man of 21 years old, he had been spurned by the Coast Guard and Navy because he was underweight. On October 27—Navy Day—1941, he entered Atlanta's Spring Street Navy Recruiting station but not before he visited the market across the street to buy and consume a bunch of bananas and a quart of milk. This time he was accepted and sent for recruit training in Norfolk.[36] Overall, more than 10 percent of the crew would hail from the Peach State.

The ship's newspaper, the *Chock "A" Block*, later observed: "Our ship being named the *Atlanta*, it is appropriate that most our crew hail from the South … of course a good percentage are from South Boston, South Brooklyn, etc., they are still southerners."[37]

One of those southern "Yankees" was William B. McKinney of New Rochelle, New York. He joined the Naval Reserve in September 1939 in the aftermath of Hitler's invasion of Poland. As a reservist, he drilled at the local armory as a member of the 31st Fleet Division and served two weeks of active duty on the destroyer *McLanahan*. In May 1941, the Navy mobilized the 31st Fleet Division and McKinney and his fellow reservists found themselves on a train to Baltimore followed by a steamboat ride down the Chesapeake to Norfolk.

Upon arrival, he recalled, "we were soon herded into a barbed wire enclosure labeled Detention Barracks." To his dismay, they were issued leggings, a garment worn by "Boots." Despite protestations, the Navy decided to treat the reservists as raw recruits.

McKinney detailed the training regimen:

> There was much close order drill, cleaning of the barracks, inspection of equipment, lecturing from the Blue Jacket's Manual and other physical exercise. With our rifles, belts and bayonets we paraded and exercised to the fine music of a United States Navy Band.[38]

After eight weeks, the 31st Division graduated and broke up, with a number of the re-minted sailors receiving assignments to attend follow-on schools. In the case of McKinney, he managed to get assigned to the Electrician's Mate School. After three months of training he graduated and spent two weeks of leave in New York. Returning to Norfolk in mid-November to await orders, McKinney "bribed" a yeoman $5 to get orders to *Atlanta*, which was being fitted out close to home. However, the yeoman arranged it that his bribe would be waived if he could round up six more Electrician School graduates who would be willing to lay down five bucks apiece under the table to get orders to the Navy's newest cruiser.[39]

As September turned to October, Ensign Graff worked topside with a chief and a first class signalman to work out the best configuration to install "blinking lights, flashing lights, signal guns, semaphore flags, and signal flags." Meanwhile, much was being accomplished decks below. Furniture and flooring were installed along with machinery and electronics. More significantly, with the hull in the water, the boilers could be lit off to generate some 665-psi steam to drive the ship's auxiliary systems such as the turbogenerators that produced electricity which flowed through miles of installed wiring. Applying steam to the turbines, the light cruiser-to-be eased down Newark Bay twice to test the propulsion and steering controls. After a month of making required adjustments and adding additional systems, it was the Navy's turn to evaluate the ship at sea. In a letter dated October 25, Adm. Royal E. Ingersoll wrote to Bowen:

> As you well know the ATLANTA represents the first of the only modern cruisers built for the U.S. Navy. Originally intended as a destroyer leader, the fluidity of tactics in the present war has suggested greater usefulness of this type of reconnaissance and new task groups. Prompt utilization of the possibilities of this type depends on full development of all her characteristics on trials. The ATLANTA will be the only ship of this type to run complete trials. For these reasons, it is requested that every effort be made to make these trials as exhaustive and complete as possible.[40]

On October 28–29, *Atlanta* left its berth and headed out to sea escorted by the destroyer *Dahlgren* (DD-187) for her initial builder's trials.[41] With the feedback received from that outing, the shipyard workers and pre-commissioning crew made further adjustments to enhance the various engineering systems. Subsequently, *Atlanta* once again headed down Newark Bay on November 21 to cross the Hudson River and up the East River under the Brooklyn Bridge for a drydocking at the New York Navy Yard—the official designation for what the locals referred to as the Brooklyn Navy Yard. After a quick hull inspection, the nearly

Atlanta during sea trials off Maine in late 1941. (Archives Branch, Naval History and Heritage Command, Washington, DC; NH 57453)

completed cruiser left on the afternoon of the 23rd for an overnight voyage to Rockland, Maine. Arriving there at 1100 to pick up the Navy's trial riders, *Atlanta* departed and began a series of full power runs of four-hour segments, astern runs of a one-hour duration, full ahead to full back "crashes," and then running tests to standardize shaft turns for 12 knots on up to full ahead. *Atlanta* developed about 80,000 SHP (shaft horsepower) and reached 34 knots on her best run. On the 27th, while en route to Boston, she steamed at 30 knots and 25 knots. After dropping off the trial riders, *Atlanta* headed back to Kearny after a brief stop at the Brooklyn Navy Yard.

In his monthly report to Bowen, Federal Shipbuilding's general manager was proud to report that *Atlanta* completed her official sea trials, "exceeding all requirements without any difficulty, thereby permitting her completion in December 1941, four months ahead of the 36 months contract time, notwithstanding a one-month delay from labor trouble in August 1941." He also noted that her fuel consumption was 10 percent below guarantees and, "with 40,000 SHP per shaft she is more powerful than any U.S. ship except one of the aircraft carriers." Finally, he boasted that *Atlanta* would be "completed ahead of time within cost estimates."[42] Plans proceeded for a Christmas Eve commissioning ceremony at the Brooklyn Navy Yard.

CHAPTER 2

America Goes to War and *Atlanta* is Commissioned

As the calendar turned to December, the future commander of this impressive warship finally arrived, having spent the previous five months completing coursework at the Naval War College. Born in 1893 in Evansville, Indiana, Samuel Power Jenkins grew up in Webb City, Missouri, attending high school there before getting a congressional appointment to attend the Naval Academy as a plebe beginning in the fall of 1911. Nicknamed "Jinks" by his fellow middies, Sam Jenkins was known for his soft step and quiet whistle. He worked hard to make it through the tough academic regimen while playing lacrosse and managing the basketball team. A handsome young man, he often attended semi-monthly hops.

To quote his 1915 *Lucky Bag* yearbook entry "sometimes he drags, more often he stags," meaning occasionally he would invite a lady friend to join him at these dances. He would be one of 10 members of his class to be tapped to join the secret "Green Bowl Society" that existed at the Naval Academy from 1909 to 1940.[1]

Midshipman Samuel P. Jenkins, Class of 1915. (*Lucky Bag*)

With war raging in Europe when he graduated, he was ordered to the battleship

Florida, which eventually joined the British Grand Fleet when America entered the war in 1917. Transferred off the battleship in August 1919, he received orders to the destroyer *Hazelwood* as the executive officer. Berthed in New York, following a Mediterranean cruise, the recently commissioned destroyer would be reassigned to San Diego, and Jenkins made the passage through the Panama Canal. Arriving in San Diego, he married Hazel Louise Hartwell on February 7, 1920. The two would spend much of the next two decades away from each other with Hazel raising their two sons as Jenkins drew back-to-back sea duty assignments.

In 1925, he checked on board the destroyer *Mullany* as the executive officer and deployed to support American operations in Nicaragua. In the early 1930s, he found himself in the Western Pacific, first commanding the gunboat *Helena*, and then *Stewart*, taking the four-piper flush-deck destroyer up the Yangtze in 1931 and 1932. In 1935, as the navigator on the battleship *Colorado*, he was promoted to commander. His first commander tour proved unique as he served as the executive officer of the *Utah*—the sister ship of *Florida* had been converted to become a target ship! Surviving that tour, he served as the pre-commissioning and first commanding officer of the destroyer *Perkins*. After a stint at the Naval Academy, in 1940 he assumed command of Destroyer Division Eight based on the West Coast. After receiving postgraduate education at the Naval War College, he traveled down to Federal Shipbuilding in Kearny as the Navy prepared to accept the ship for commissioning. The "Old Man" was 48 years old when he took command.[2]

The prospective CO's arrival coincided with the return of Ensign Corboy from gunnery school in Washington. On the following Sunday, Corboy went to Radio City Music Hall to see the Rockettes and the annual Christmas show. Little did he realize, as he watched the holiday performance, that hundreds of Americans were losing their lives on an island in the mid-Pacific. When he emerged from the great hall, he noticed that crowds had gathered to watch the moving news bulletin on the building. Then he saw the headlines on freshly printed newspapers that newsboys were briskly selling.

Van Ostrand and Jane "Betts" Perkins were spending the weekend with their parents in their native Greenwich, Connecticut. They attended a

Christmas dinner on Saturday night, and on Sunday, the couple took a morning walk in the woods before attending a luncheon with their two families. A family friend, Jeanie Mullin, prepared the meal and brought in the entrées from the kitchen. As the two families enjoyed the good food and conversation, Jeanie came bursting in: "They just bombed a precious stone! They just bombed a precious stone!"

All rushed to the family radio where they heard the announcement "Japan has just bombed Pearl Harbor. All officers and men will report to their ships and stations in uniform immediately." Perkins rushed upstairs to his old room to find a uniform. Finding only an old midshipman dress hat, he donned that on his head, packed, took his wife and their pet dog "Bini," and sped off to New York with the two families standing in the driveway looking on.

Speeding down the Merritt Parkway at 75 mph, a motorcycle cop noted the hat and sped ahead with siren running to clear the way. After they briefly pulled over to tell the cop where they were heading, the two-vehicle motorcade continued, and Betts asked an obvious question: "What are we doing careering along this to a ship that's not going anywhere? Can't we slow down?" To which her husband responded: "I wouldn't deprive this nice young man of the chance to do something for his country … Besides, this is the first, and presumably the last time in my life I can go 30 miles an hour over the speed limit, shoot the red lights, and break every law in the books with the aid of 'the arm of the law.'"

The motorcycle escort took them to their Greenwich Village apartment where Perkins changed into his full uniform. The cop watched over the car which was double-parked. Betts then drove her husband through an eerily empty lower Manhattan, on through the Holland Tunnel over to Kearny to the gate of Federal Shipbuilding. Depositing her husband, she drove back to the apartment.

Reporting for duty, Perkins was told that since the ship was not going anywhere, the duty section had things well in hand, and he could return home. In contrast, Ensign Graff, who heard the breaking news during an interruption of a broadcast of the New York Philharmonic while sitting fireside with his parents in Scarsdale, simply made a phone call to the ship's duty number and was told there was no need to come down.[3]

The following morning Captain Jenkins gathered the pre-commissioning crew of officers. chiefs, and junior petty officers in the ship's office to listen to President Roosevelt's call for Congress to declare war against the Empire of Japan and the subsequent action by Congress to do so.

Shocked by the Japanese attack, Rear Admiral Bowen expressed disappointment upon hearing Army radar had detected the incoming aerial assault, but warnings went unheeded. His team at the Naval Research Laboratory had pushed hard to field the new technology, and the Pearl Harbor outcome proved demoralizing. Given the outbreak of war, he yearned to return to Washington to focus his attention on developing further technologies that could lead to victory.[4]

Fortunately for Bowen, the union would help to facilitate that quest. Local 16 President Peter Flynn sent Bowen a sincere supportive letter which enclosed the following resolution:

> Whereas: The acts of treachery of the Japanese Government have amply demonstrated the desire of the Axis combination to enslave the world with a chain of bondage conceived in the diseased minds of Hitler cohorts.
>
> Be it resolved: That we, the members of Local #16 IUMSWA—CIO pledge our wholehearted cooperation to the President and the Government of the United States, in the crusade for the preservation of democracy in a world which is being harried and destroyed by the "mad dogs" of the Axis, and may it be further RESOLVED that we, the officers and members of Local #16 pledge ourselves to land every effort to increase our production of the vital defense vessels so necessary for the extermination of the menace which at this moment threatens the security and freedom of the democratic peoples of the world.[5]

With Flynn's missive, the rear admiral took advantage of the opportunity to extract the Navy from running a private shipyard. Bowen recommended to Knox to return the shipyard "while the going was good." The Navy Secretary agreed. On January 5, 1942, President Roosevelt signed an executive order relinquishing Navy control of Federal Shipbuilding and Dry Dock Company.[6]

With America declaring war not only on Japan but against Germany and Italy, a sense of urgency existed to get *Atlanta* to sea. The Navy needed combatants to replace the battleships lost at Pearl Harbor and then some.

Betts Perkins noticed her husband was now arriving home from Kearny exhausted. "It seemed he was constantly plagued about the plumbing, and in particular the captain's 'head.'"

Prior to the attack on Pearl Harbor, Betts had been making plans to host a party for the *Atlanta*'s officers and their wives prior to its commissioning. In the wake of the attack, Van Ostrand told her to continue planning, as he knew the preparations would keep her mind off his pending combat. Flowers filled all the vases of the Perkins apartment as the officers and their spouses arrived one December evening for a buffet dinner. One ensign thoughtfully brought more flowers. Betts gave them to her husband with instructions to put them in water and then she proceeded to greet the arriving Captain Jenkins and his wife. The *Atlanta*'s prospective skipper impressed Betts. She saw him as "a gracious, kind, and thoroughly capable, all the qualities of a fine leader." Over two dozen people crammed into the apartment to join in on the meal and the after-dinner coffee. During the coffee, Captain Jenkins excused himself for the bathroom. He quickly returned. Perkins asked if there was a problem. "Well there seems to be a large flower arrangement in that part of your bathroom that is most interesting to me at this moment." Amid the laughter, Perkins took care of the obstruction. The captain looked at Perkins again, "You wish all your plumbing were that easy and smelled as fresh, don't you young man?" Betts observed that despite the flower incident, or because of it, the dinner was a success: "We had become a family—the men liked one another, the wives were good friends, and we all took pride in the ship."[7]

At dawn on December 23, *Atlanta* departed the shipyard to steam down the Hackensack one final time into Newark Bay and then turned left to pass under the Bayonne Bridge, the world's longest arch span, past the eyes of the Statue of Liberty, then across Upper New York Harbor Bay to the East River, passing under the Brooklyn and Manhattan Bridges before arriving at pier G, berth 12 at the Brooklyn Navy Yard.

With rain predicted for Christmas Eve, a tarpaulin was rigged over the ship's quarterdeck and stern section to keep the expected dignitaries dry during the commissioning ceremonies. By 1100, the official party took their seats on the quarterdeck, overlooking the seated guests on the pier. The ship's crew, which had grown to over 200, stood in formation with the officers lined up in ranks amidships facing forward, and the enlisted crew standing in ranks to port and starboard.

Many of the enlisted sailors were seeing their ship for the first time. Electrician's Mate McKinney arrived in New York on December 5 and checked into pier 90 on the Hudson at 50th Street. The Navy had converted the warehouse structure into a giant bunkhouse and berthed the old cruiser *Seattle* alongside to provide messing facilities. He spent the weekend "on the town" learning of the attack on Pearl Harbor only late on Sunday. After the attack, he was sent down to the new receiving barracks built across from Navy Yard, New York—also called the Brooklyn Navy Yard. Issued a gun, he was posted at a nearby Subway kiosk. "Apparently the Japs were expected to arrive on the BMT," he observed.[8]

The official party included Third Naval District Commandant Rear Adm. Adolphus Andrews and Rear Admiral Marquart.[9] The Captain of the Yard, Capt. Harold V. McKittrick read the commissioning orders. With that Captain Jenkins directed his crew to face aft and called them to attention as two sailors unfurled the national ensign. Following the playing of the national anthem, Jenkins read his orders assigning him as commanding officer. Looking out at the assembled crowd, he stated: "It's an honor indeed to be the captain of the first major warship to join the Navy since Pearl Harbor."[10] Completing his prepared remarks, Jenkins then set the first watch. The chief boatswain's mate stepped forward and piped the call, and the bluejackets broke ranks to hustle to their assigned battle stations. On the quarterdeck, Lt. Cdr. James S. Smith assumed responsibilities as the first in port officer of the deck. Admiral Andrews gave a blustery speech stating, "this is a fighting ship" and reminded those embarked: "your job from this day forward is to use its guns to blast and smash the enemies of this country."

The ship's sponsor, Miss Mitchell, then addressed the crew of the ship named for her city. Wearing the blue-gray uniform of an American Red Cross canteen volunteer, she told *Atlanta*'s plank owners, "We know that when the time comes you will give a wonderful account of yourselves." Due to Mitchell's fame, the City of Atlanta commissioned a portrait of

The commissioning ceremony held on Christmas Eve at the Brooklyn Navy Yard followed by the presentation of a silver bowl by Margaret Mitchell to Captain Jenkins. (Archives Branch, Naval History and Heritage Command, Washington, DC; NH 57450; NH 57449)

the author instead of a silver set. She also presented Captain Jenkins with a large silver punch bowl.

The skies cleared as the speeches went on, and as the last words were spoken, America's newest warship basked in sunlight. Ensign Corboy observed, "A rather dull tableau suddenly was a scene of splendor." A wife of one officer later wrote: "To my artist's eye she was a thing of beauty and a true oceanic lady."[11]

McKinney described *Atlanta* as "the most beautiful ship that had ever been built." H. Charles Dahn, a 17-year-old who had just graduated boot camp in Newport, Rhode Island, recalled, "From the first moment I saw her in the fog-shrouded Brooklyn Navy Yard, I knew that she was destined to do great things. With her clean swept lines, she loomed out of the fog like a long, gray ghost."[12]

After the enlisted crew secured from their battle stations, they were joined by their families and friends on the ship's mess decks for a hearty lunch while the officers and their kin were served a meal in the ward-room. After 1600 the officers departed for a more elaborate reception at Manhattan's St. Regis Hotel that was paid for by Cola-Cola. Bob

Graff recalled it was a marvelous Christmas Eve party that made it "very difficult to get home."[13]

Once on board the new crew found the accommodations to their liking. Looking at his new bunk and locker, McKinney observed, "The Navy was getting better." However, he did not have time to get comfortable as two first class petty officers put him right to work setting up the ship-to-shore telephone switchboard.

McKinney worked through Christmas Eve and finally obtained liberty at noon on Christmas Day in time to go home to join his family for a holiday feast. Not all got to go home that day since a duty section needed to remain aboard to assure the ship remained safe from fire or flooding, and to handle the arrival of provisions, such as 60 gallons of milk from the Borden Dairy Company which arrived that morning. For those with homes in the New York area, this Christmas was a bit emotional as families and wives suspected it would be the last with their loved one for years to come. Betts recalled she went "hog-wild" purchasing gifts for her husband, including a velvet smoking jacket, humidor, pipe, and tobacco. She also purchased a white fluffy toy bunny from FAO Schwartz as a good luck talisman for Van Ostrand Perkins to take to war.[14]

Holiday aside, there would be no lightening of the workload as the crew familiarized themselves with their new home. "I learned where to be, when, and why, as well as what to do when I got there," recalled McKinney. He was assigned to "E" Division, a group of 25 electrician's mates led by a new division officer and two chiefs. That division officer was Ens. Walter M. Straub, Naval Academy Class of '42. His commissioning had been rushed by six months and along with Ens. Don Spangler who came on to be the radio officer. McKinney would also become familiar with Ens. Dave Nicholson, who checked in shortly after Christmas, having graduated top of his class of a reserve commissioning program based in Annapolis for men with engineering backgrounds. Given the quality of talent and the importance of electricity in all facets of shipboard operation, McKinney quickly discerned that his was the most important division on the whole ship. Other crewmembers began to gain a similar appreciation for their divisions.

Just out of boot camp in Norfolk, Clifford Dunaway was part of the Georgia contingent that would constitute about 10 percent of the crew. Having reported aboard just in time for the commissioning ceremonies, he stood at morning muster one day and raised his hand in response for a call for gun strikers. By volunteering Dunaway signaled his interest in "striking" for the gunner's mate rating. Dunaway would be assigned aft to turret eight, the furthermost aft 5-inch 38 twin mount where he would be assigned as a loader. Though the work entailed in maintaining and cleaning the guns was hard, Dunaway took pride in the job which kept him out of deck swabbing or mess cooking duties.[15]

Having also gone through basic training at Norfolk, Franklyn LeRoy Reed could also claim he had southern roots—if southern Connecticut counted. A native of Mystic, Reed had an interest in the Navy since his father worked for the Electric Boat Company and he had been an avid sailor on Long Island Sound.[16]

To train recent recruits such as Dunaway and Reed, a strong core of seasoned chiefs and petty officers were assigned. Boatswain's Mate First Class Leighton Spadone had joined the Navy during the middle of the Depression in 1935 and had spent six years assigned to the battleship *California* prior to coming to *Atlanta* to serve as the leading petty officer for Fourth Division, which would have responsibilities for the 1.1-inch and 20mm guns positioned around the forward deckhouse. Spadone would later write: "My very best memories of 24 years in service were in the *Atlanta*."[17]

Not everyone would adjust to life in close quarters with their fellow shipmates. The deck log entry for the 4 to 8 watch on December 29 recorded a fight at 0500 between an abusive Seaman First Class Slabos and Seaman First Class Campbell. The fight was broken up following Campbell stabbing Slabos in the chest, causing a slight laceration. Both men were placed in P.A.L. status, meaning Prisoners at Large—in essence, confined to the ship until further notice. Following a very short underway period where *Atlanta* was shifted to pier C, berth 3, Captain Jenkins held his first Captain's Mast where he had authority to mete out non-judicial punishment for minor infractions. For sailors brought before him who had been late returning from liberty, reported late to stand a watch, or were overheard to use foul language towards another shipmate, Jenkins established a precedent by either withholding a day or two of liberty or assigning the charged individual extra duty at the end of the day for so many days. For more serious offenses such as multiple days of absence without leave (AWOL) or the type of incident that had occurred earlier, Jenkins had the authority to convene a summary court-martial or a deck court. A deck court consisted of one commissioned officer who would be given authority to dismiss charges or impose punishment on the accused. The prosecutor, known in naval legal terminology as the recorder, would present the government's case, calling forward witnesses, and introducing evidence as appropriate. The accused could ask for an individual to represent him and that individual need not be a commissioned officer. The accused could also decline to appear before a deck court which was a calculated gamble in that Captain Jenkins could then convene a more elaborate summary court-martial which had authority to mete out harsher punishments.[18]

In the case of Slabos and Campbell, Jenkins opted to convene a summary court-martial, a tribunal of commissioned officers with the highest-ranking officer appointed as senior member. In this case and in the majority of the future summary courts to be convened in *Atlanta*, the Navigator Lt. Cdr. James S. Smith, would be appointed as the senior member. That Smith would be the commanding officer's go-to guy for summary courts likely reflected that the ship's navigator's in-port chores consisted mainly of acquiring proper charts and making sure they were being kept up to date. In the ship's collateral duty bill posted on May 31, 1942, Smith was listed along with Lieutenant Commander Wulff and Ensign Spangler as standing members of the court with Ens. Ira W. Wilson appointed as the recorder.

Smith would convene his tribunal just after the new year and Slabos would be found guilty of "using threatening language towards another member of the naval service" and Crawford would be found guilty of assault. Unlike a deck court where Captain Jenkins as convening authority could direct immediate imposition of punishment, summary courts finding had to be forwarded to Jenkins's immediate superior for review and approval. In the case of Slabos, the immediate superior approved five days of confinement with a subsistence of bread and water with allowance for a full ration on the third day plus a deduction of $20 from his pay for the following month. Crawford earned 10 days of confinement on a bread and water diet with a full ration permitted on the third, sixth, and ninth days plus a pay loss of $20 for two months.[19]

A tradition, that came into being two decades prior, called for the first deck log entry penned by the first watch of the new year be in rhyme. One stanza of *Atlanta*'s entry read:

> To comfort our ship, ere she goes forth;
> Be it Tropic Clime or the frozen north;
> To blast our foes from their evil thrones;
> We receive from the dock steam, juice, water, and telephones.

Throughout January the crew found themselves, according to Ens. Ed Corboy, "working high speed, early and late" to tackle the ship's many deficiencies.[20] On January 9, shore connections and lines were disconnected to allow for *Atlanta* to be placed into dry dock #4. For the next 11 days the light cruiser rested on keel blocks as yard workers and ship's company confronted issues within the engineering plant, interior communications, and electronics.[21]

Although safely ensconced at the Navy Yard, the reality of war was ever present in the newspapers and radio reports that tracked the rapid expansion of the Japanese Empire and the advances of German troops to the outskirts of Leningrad and Moscow. Often the crew witnessed merchant ships pulling into the harbor with crushed in side plates from torpedoes. Undergoing battle damage repair was HMS *Dido*, a British cruiser heavily damaged in the Mediterranean. Shipyard workers claimed of entering several of the cruiser's compartments to find body remnants still present.[22]

On February 8 at 0912, tugs pulled the light cruiser out into the East River as lines formerly attached to bollards at pier C, berth 3 were pulled aboard by *Atlanta*'s deckhands. With a pilot recommending speed and rudder orders, the ship passed under the Manhattan and Brooklyn Bridges. One sailor topside looked aft and observed "her wake resembled that of a speeding Chris-Craft on a quiet lake. The water immediately astern of her churned furiously, and the waves stretched to either bank." With a pilot boat coming alongside to retrieve the pilot, as the light cruiser passed Lady Liberty, Captain Jenkins assumed the conn and guided *Atlanta* through the Verrazano Narrows and ordered the helmsman to steer towards an anchorage spot in Gravesend Bay on the Brooklyn waterfront just northwest of Coney Island. With the ship's anchor settling below at 1012, the command "shift colors" was broadcast over the 1MC—the general announcing system—and simultaneously national colors were lowered from the forward mast and broken out on a flagpole off the stern. Simultaneously, a blue-and-white union jack, containing 48 stars representing the states of the republic, was raised up on a flagpole at the bow.[23]

Atlanta anchored at Gravesend Bay to take on ammunition delivered by a barge that came alongside at 1250. Working parties hustled all afternoon, into the night and into the following day to bring aboard 5-inch shells, powder cartridges, and smaller caliber ammunition onto the ship. A detailed inventory of the assortment of ordnance hauled aboard was written into the deck log, filling two complete pages.[24] Bob Graff recalled that each division was assigned to provide so many bodies to work in each party and that a junior officer would be tasked to supervise—and quite often that fell on him! Graff saw this as a team-building exercise noting an inclination in some to goof off: "I gathered them together and said if you don't lift the case, he has to lift two cases … and pretty soon he's going to be very mad at you and he'll probably take a swing at you." Graff was not shy about lending a hand, clambering down into the barge to help sling cases and that evoked a response: "As soon as they were convinced their officer was for real, they turned in a performance."[25] To keep spirits up, the ship's band rehearsed a variety of tunes. While the 5-inch shells and powder cartridges slowly filled up the forward and aft magazines deep below, sailors stowed 20mm and 1.1-inch munitions in small ready service rooms dubbed "clipping rooms" placed near those gun mounts.

Following the midday meal on February 10, tugs arrived to remove the barge, and by 1459 the anchor cleared the mud below. "Underway,

Shift Colors" was announced as Captain Jenkins took the conn to issue rudder and engine orders to the helm and lee helmsmen. As *Atlanta's* shafts began to churn, the now-armed cruiser worked through open netting and steamed by the Ambrose Lightship. Below, crewmembers continued to stack the projectiles in the magazines, greasing the casings and inserting tracer rounds. With the cruiser clearing Ambrose channel, the captain ordered General Quarters. Over the next 42 minutes, the crew remained at their battle stations as "manned and ready" reports filtered their way to the bridge. With the ever-present U-boat threat, Jenkins ordered the lee helm to shift the engine order telegraph to full for both shafts, and began a series of course changes with the implementation of a zigzag plan, as the ship proceeded on an overnight voyage to Norfolk. Overhead a blimp based out of the Lakehurst Naval Air Station patrolled to spot any German submersibles. With the arrival of darkness, the airship departed, and *Atlanta* continued ahead under darkened-ship conditions. For the underway officer of the deck duties, Lt. Philip T. Smith and Lt. Van Ostrand Perkins took turns up on the bridge.[26]

With the departure from Brooklyn, the crew began to adjust to the "at sea" routine of standing watches, performing maintenance and chores, and conducting training. While in Brooklyn many of the officers and crew who were not assigned to stay with the ship as part of the duty section, left for their homes or rented apartments. Bob Graff continued to commute daily from Scarsdale. "Officers Country" in *Atlanta* was located one deck below the main deck, located in the forward section of the ship. The senior department heads had their own staterooms. Lieutenants typically shared a stateroom, occasionally flipping a coin to call the top or bottom bunk. Further forward the ensigns and junior grade lieutenants stayed in rooms having either three or four sets of bunk beds. Graff recalled the rooms were painted a pale apple green "because psychological studies that the navy made over the years told them that green was the most neutral color."

The wardroom was up on the main deck located two decks below the bridge. Bob Graff recalled the officers' dining room extended across the width of the ship minus the exterior passageways. An adjacent small galley supported the officers' mess with food offerings, but Graff recalled most of the food was brought up from the main galley: "Then the Filipino stewards dealt with it, maybe fancied it up a bit."

Whereas Captain Jenkins had a dining table in his cabin, his own cook, steward, and ate alone, Commander Emory was "the Lord of the Manor" in *Atlanta*'s wardroom. His chair was at the head of the table on the starboard side, and he was flanked by his department heads. Lieutenants filled the far end of that table. The remaining lieutenants and jgs sat at the midships table, and the most junior officers took their places at the port side table. The stewards were instructed on who had seniority and placed the napkins with engraved napkin holders accordingly. All stood behind their chairs awaiting the arrival of the XO. When Emory sat down, the rest of the officers followed his lead. If for some reason an officer needed to depart early to stand a watch or tend to an immediate duty, he would approach the XO and request permission for dismissal.

Considering the carnage that was occurring at various locations on the seven seas, Graff found the meals a very civilized experience. For breakfast the officers coming up from their staterooms or coming off watch would go to the bulkhead counter to pour their tea or coffee and grab a glass of juice. They would then sit down, and the stewards would serve them a choice of morning meal offerings. Graff recalled the noon and evening meals as more elaborate, with two or three course offerings opening with soup, followed by a main course, and concluding with dessert of either fresh or canned fruit. The stewards brought the food out on silver-plated pewter platters. Nobody would touch his food until all at the table had been served.

One unique aspect about wardroom dining is that the officers were billed for their meals monthly as the Navy provided a meal allowance with their pay. Ensign Wilson had the dubious honor of taking on the role of mess treasurer, using collected funds to pay for food drawn from below decks or from off ship markets when those opportunities presented themselves.

For the enlisted, most slept in larger bays further below deck. In the case of McKinney, he was assigned a "rack" in a bay located two decks below Officers' Country. As additional crewmembers arrived, eventually hammocks were installed to accommodate the excess hands. Unlike the officers, the enlisted crewman would grab a metal tray and run through a serving line. Graff recalled that some of the sailors, emerging from areas of the country where the effects of the Great Depression still took a toll in the dining room, the serving of meat every day caused indigestion problems. Overall, the bluejackets serving in *Atlanta* had few complaints about the chow.[27]

With choppy seas bobbing the light cruiser along, many of the new sailors felt nauseous. The next morning at 0630, the boatswain's mate of the watch sounded General Quarters, a ritual that continued at dawn and sunset at sea when the light cruiser was most vulnerable to being ambushed by enemy submarines. After securing from GQ at 0742, *Atlanta* rendezvoused with a pilot boat to take on a pilot as the ship held the Cape Henry lighthouse off the port beam. Under the pilot's guidance, *Atlanta* maneuvered up the Thimble Shoals channel, into the Chesapeake Bay, and into Hampton Roads, anchoring at the assigned berth 21 at 1411. Shortly thereafter, a small boat came alongside carrying two physicists and a radio engineer from the Naval Research Laboratory.[28]

The next morning *Atlanta* was once again underway as the ship departed Hampton Roads to exit Thimble Shoals channel for operations off Cape Charles. German magnetic mines exploited a natural occurrence as ships' hulls became magnetized as they steamed over the planet's magnetic fields. To eliminate this magnetic signature, underwater degaussing ranges were laid out nearby naval bases in the early 1940s to neutralize a ship's magnetic field. Arriving at the head of a recently installed range at Cape Charles, with the Naval Research Laboratory personnel embarked, *Atlanta* made repeated runs along the range—back and forth a total of 17 times as underwater sensors measured the ship's magnetic footprint. Upon finishing the degaussing runs at 1730, Captain Jenkins turned up into Chesapeake Bay and anchored at 1847 off Nassawadox Creek along the bay's eastern shore.[29]

Underway the following day at 0748, *Atlanta* continued to ply in waters protected from U-boats but not from bitterly cold winds. Friday the 13th saw *Atlanta* steam in endless circles as the Navigator, Lieutenant Commander Smith, and his quartermasters compared readings off the gyro compass and the magnetic compasses to determine the deviation on the latter following the degaussing. Additional tests were conducted by the visiting physicists to calibrate *Atlanta*'s radio detection finding gear. Once the maneuvers were completed, the light cruiser turned up the Potomac River and arrived and anchored that afternoon off Mattawoman Creek adjacent to the Indian Head Naval Powder Factory, presumably to drop off the Naval Research Laboratory physicists.[30]

On Valentine's Day, *Atlanta* steamed back into Hampton Roads. After anchoring at berth 21 before noon, the light cruiser took on 52,849 gallons of black oil from a fuel barge to replenish her bunkers. At Captain's Mast that afternoon Captain Jenkins meted out warnings and extra duty sentences to enlisted sailors who had attempted to evade censorship restrictions in outgoing correspondence. One of the unique collateral duties as ship's censor fell on Lt. Philip T. Smith. He and three other junior officers screened letters before they left the ship.[31] The wife of Van Ostrand Perkins would later write that the officers skirted the rules by employing a series of codes to communicate their whereabouts to their spouses:

> We set up a code so Van could let me know where he would be. Mention Audie or Mrs. D and the ship was in Honolulu. A. Mabani (our apartment in Manila) meant the far Pacific. Willow Blathly, (Van's English ex-girlfriend) meant England. The Raleigh (his old ship) meant the Mediterranean and Elvira's Farm meant San Francisco. As for the dates we used a fictitious name. Starting with the month's first three letters and saying the day thus—"Martha's birthday is the sixteenth, please send her a card"—meant March 16th.[32]

The next two days saw *Atlanta* again underway, primarily to calibrate the radio detection finding gear. Reception of wavelengths of 485 kilocycles were followed by 380 kilocycles and so on. The rest of the crew hardly stood idle as the First Lieutenant, Lieutenant Commander Sears, and Lieutenant Perkins challenged the crew to complete an exhausting series of damage control drills to simulate flooding, fire, and collision. In addition to developing practicing firefighting, bulkhead shoring, and dewatering skills, the crew listened to first aid lectures and attempted hands-on training to set splints, apply bandages, and administer morphine.

Following an overnight anchorage in the Chesapeake on the evening of the 16th, *Atlanta*'s gunnery department swung into action. Lt. Cdr. Bill Nickelson, aided by Lieutenant Mustin and the division officers, calibrated the guns and radars, and began training their prospective gun crews. Nickelson would oversee operation of the forward 5-inch mounts and Mustin took command of the aft mounts. Gunnery remained labor intensive. Ensign Corboy, whose Third Division had responsibility to man up the aft three 5-inch 38 gun mounts, had 118 men assigned to work in the mounts, handling rooms, and magazines below. The newly formed gun crews slowly rehearsed loading and firing the guns for the first time. Each 5-inch 38 mount had a crew of 13. Working quickly together, a crew could discharge a salvo every four seconds. However, to maintain that pace, shipmates below in the upper handling rooms had to rapidly feed shells from the magazines into the electro-hydraulic hoists. The inaugural test shots occurred on the afternoon of February 17, from the 5-inch 38 twin mounts totaling 64 rounds. That broke down to four rounds per barrel. In addition to breaking in the new gun crews, such test shots were needed to determine whether structural damage might be caused, or interior communications disrupted.[33]

The test shots did cause several of the phone circuits to go offline. Consequently, Petty Officer McKinney spent much of the period topside, hooking up and repairing the sound-powered phone system. In principle, the sound-powered phone system worked like two cans connected with a string, a system familiar to grade school students. This system was a bit more sophisticated, using headsets that had mouthpiece attachments. The speaker would depress a button and speak into the mouthpiece, vibrating a diaphragm, moving a tiny armature in a coil, "thus setting up an electromotive force, which traveled along the conductor or connecting wire and energized the coil in the receiving earpiece." The circuits could be cross connected through a switchboard found in the I.C. (Interior Communications) room. Because of the dexterity needed, McKinney hooked up the lines with his bare hands until they were so numb he could no longer hold tools. "The wind cut like a proverbial knife," he recalled.[34]

Following another evening anchored in the Chesapeake Bay, *Atlanta* got underway at 0749 on February 18, reporting to the commander in chief, Atlantic Fleet, for duty. That day gun crews broke in the smaller 1.1-inch machine guns and on the following day some 480 rounds were fired from the 20mm guns for the first time. Underway at 0754 on Saturday February 21, the crew went to General Quarters to conduct a series of drills to include the transfer of ship control to Battle II, the backup bridge located just aft of the second Mark 37 gun director. From Battle II, the Executive Officer, Commander Emory, assumed control of the rudder and passed down orders to the engine rooms to change speeds. Petty Officer McKinney would later write, "Speed runs, maneuvering, fire drills, fire and rescue party, collision, abandon ship, man overboard—we learned them all."[35]

After a grueling week, *Atlanta* returned to Hampton Roads to anchor at berth 25. Lookouts took note of the carrier *Hornet*, which was readying to depart for the Pacific. After dispatching a contingent for shore patrol duties, the announcement of "Liberty Call" was welcomed by many of the crew.[36]

Those going ashore braved a half-hour boat ride. "The trip ashore in a motor launch can be a wild experience in rough weather," recalled Petty Officer McKinney. McKinney and three more senior shipmates checked out a bar on Norfolk's East Main Street. Seated in a booth, the four sailors took swigs from tucked-away liquor bottles that had been slipped to them by the waitresses since Virginia law forbade hard liquor being served in such establishments—only beer and wine were served. Sitting across from one heavily tattooed shipmate from Wyoming, McKinney learned about life with the Asiatic Fleet.[37]

Also awaiting ashore were a half dozen of the officers' wives. While Hazel Jenkins and Betts Perkins got to spend a few hours with their respective spouses, the other four women would return to New York disappointed that their husbands could not make it off the ship that evening.[38]

The liberty call proved to be a one-night stand. Sunday morning again found *Atlanta* underway to return to the Chesapeake to anchor off Tangier Island that evening. Next morning while underway, the crew conducted on-station training following the sounding of General Quarters and then secured from battle stations for the midday meal. Just past noon the crew again ran to their battle stations. The gun mounts came alive, starting with the forward mount one unleashing 5-inch rounds and working on back to the aft mount eight where Seaman Dunaway patiently waited for his chance to load shells. The deck log recorded 127 rounds expended. During the exercise, the gun crews learned to deal with adversity, handling two hangfire incidents where shells had to be extracted and tossed over the side after failing to fire.[39]

After another night off Tangier Island, *Atlanta* again raised her hook just before 0800 on Tuesday February 24, and once again the General Quarters alarm was sounded. Nickelson's and Mustin's 5-inch 38 gun crews eagerly awaited the arrival of a plane towing a target sleeve. On the aircraft's first pass, the light cruiser fired four rounds at the sleeve. On the second pass following six shots, the sleeve parted and fluttered down into the bay. With no target sleeve, the 1.1-inch and 20mm gun crews took turns firing at balloons released into the wind. Mustin noted that most of the balloons flew away unscathed. This routine continued over the rest of the week.

What frustrated *Atlanta*'s gunners was employing the fire-control radar to target the ordnance. The primitive FD radar required teamwork to operate. The "trainer" had the chore of sweeping the antenna towards incoming targets that were reported by the long-range search radar or lookouts. Once the trainer determined the approach axis and vertical angle, a radar range operator took charge and provided an estimated distance for the plotting room. In addition, a sailor was assigned to an optical range finder to examine visible targets to cross-check their identity. Over the Chesapeake that February, aircraft pilots towing the target sleeves depended on those optical range finder operators to keep them alive.

With two weeks having passed since departure from Brooklyn, a small floatplane landed aside of the light cruiser on the on the morning of the 26th to deliver and take outgoing mail, providing a morale boost for many in Jenkins's crew. Writing passionately to his wife, Van Ostrand Perkins penned:

> I miss you like fury Bettsy girl. More and more in every way I love you and the fact that I am not with you is very disappointing. Betts, my own, this is turning into one voyage to which I have not and cannot give my heart. Regardless of where I am or where I go all of me that remains any good remains with you.[40]

Later that day Jenkins would again have to maneuver his ship to pick up a fallen sleeve. Following early gunnery drills the next day, *Atlanta*'s topside sailors witnessed two of America's newest battleships as *North Carolina* steamed by to port and *Washington* passed to starboard. With the two new battlewagons now astern, Captain Jenkins conducted a man-overboard drill. For the newly minted line officers needing to work on their shiphandling skills, the "Man-Overboard" drills offered good opportunities. While the junior officers took turns maneuvering the ship to retrieve the dummy tossed over the side, the crew mustered to determine who was missing. To inject realism, the drill coordinator would grab a crewman and hide him in a closet. In addition, the crew learned where to muster should there be the need to abandon ship. During these drills it was determined that the placement of stacks of life rafts on the stern of the ship made quick abandonment of the ship impractical.[41]

That Friday afternoon, *Atlanta* again traversed Thimble Shoals channel and anchored at berth four at 1458. Shortly thereafter, Captain Jenkins sent a contingent of shore patrol ashore and crewmembers not standing duty once again took advantage of motor launches to spend some time ashore. Nearly 25 hours after arriving, *Atlanta* again raised her hook and steamed the outbound passage through Thimble Shoals channel and anchored at Lynnhaven Roads near Cape Henry. On Sunday, the engineering department became the center of focus as the light cruiser conducted full-power runs. Lieutenant Commander Loeser oversaw the runs to determine the number of rotations per minute needed to attain each additional knot of speed. As different speeds were ordered, acceleration and deceleration rates were measured. Finally, the ship underwent flank speed runs, both ahead and astern. As *Atlanta* approached the designed top speed, crewmembers aft began feeling excessive vibration. Bob Graff would recall: "... the vibration was so great that crockery broke, and the gunnery sights were unusable, and it was just mayhem.[42]

Over the next two days, *Atlanta*'s crew ran through additional gunnery and damage control drills with a new wrinkle added late on March 3 as Captain Jenkins took his crew on a nighttime cruise in the Chesapeake to test their ability to spot targets in the wake of the recent Battles of Java Sea and Sunda Strait where the Japanese Navy twice bested a combined flotilla of Australian, British, Dutch, and American warships. The loss early on March 1 of the heavy cruiser *Houston*, along with the Australian cruiser *Perth*, meant the prewar American Asiatic Fleet had been effectively eliminated.[43]

The next morning, Commander Emory took the conn as *Atlanta* conducted gunnery exercises and briefly met up with *North Carolina*. Then the light cruiser steamed back into Hampton Roads and directed to berth port side to pier 7 to receive additional ammunition and supplies. Of note, the Supply Officer, Lieutenant Grubbs, went ashore to have $31,000 dollars signed over to him to cover the forthcoming payday. After spending most of the next day, March 5, pierside, *Atlanta* departed for a short trip into Hampton Roads to again drop anchor. The following morning on March 6, *Atlanta* departed for Portland, Maine.[44]

En route, *Atlanta's* sailors witnessed combat for the first time as the lookouts spotted two destroyers on the horizon firing their guns. Captain Jenkins aimed the bow of the ship in the direction of the conflagration and sent the crew to General Quarters. Upon arriving on scene, Jenkins found the two tin cans moving across waters smeared with oil and debris. One destroyer signaled that they had just sunk a sub and they were searching for survivors.[45]

Steaming east of Cape Cod, *Atlanta* surged forth into high winds and heavy seas. With green water breaking over the bow, the weight of the sea pounding down on the open steel deck caused the ship to shudder. Having steamed for three weeks in the protected waters of the Chesapeake, the crew had to adjust to 25-degree rolls that tossed loose gear and weakened even the stomachs of Old Salts. Up on the bridge, the watch teams which rotated between Lieutenants Perkins, Smith, Mustin, and Wulff, endured a heck of a ride. To accommodate the seasickness, a bucket was strategically located on the bridge. The smell of vomit reeking through the pilothouse only worsened the nausea experienced by those on watch. For Merrill Peyton, assigned to First Division, the topside watches proved unbearable. He requested and received permission to be transferred to the engineering department.[46]

On the afternoon of March 7, land was sighted and the navigator scanned the horizon for landmarks that would confirm *Atlanta*'s location. Coming up on deck, Petty Officer McKinney recalled it was clear and cold and his love for sea air was further enhanced. Looking ahead at the horizon he would later note, "My first look at Maine's rock-bound coast stamped it as a part of the country to be forever remembered with great favor." The cruiser passed Cape Elizabeth and into Casco Bay.[47]

Tipped off about the ship's destination before the *Atlanta* left Norfolk, five junior officer wives took a train up to Boston, only to find a connecting train would not depart till the following afternoon. Not to be deterred, Jane Shaw, Stephanie Wulff, Ellie Broughton, Helen Hall, and Betts Perkins hired a taxicab to take them along the Maine coast to their destination. Arriving in the middle of the night, the women awoke the night clerk in the lobby of the run-down Eastland Hotel. On the clerk's recommendation, they took rooms on the top floor to take advantage of the rising heat in the building.

Atlanta underwent additional training over the next week. Bob Graff recalled there was extensive cross-training. Not only did his sailors work on their communication capabilities, they would also go down to one of the 5-inch 38 mounts and be told they would have to shoot the gun within an hour and they had to figure it out: "One guy would say 'I'll take the elevation,' and another guy would say 'I'll take the horizontal,' and another guy would say 'I'll take the loaders position' and so forth." Graff noted his men were trained on torpedo tubes and damage control: "They were even put in the engine room which God knows was Oz to the rest of us." With the chief engineer watching, Graff's men learned what valves needed to be turned in an emergency.[48]

In contrast to Norfolk, liberty was allowed most nights with half the crew being allowed to go ashore. During the day the women would hang out together at the hotel since there was not much to do in Portland.

During the evenings, they would go out with their husbands for cheap lobster dinners and then get cozy with them in their hotel rooms. The lucky hubbies often got off a duty section by making arrangements with other officers to cover for them.[49]

For the enlisted, Portland didn't offer much. Several sailors took advantage of "Church Call" to attend services on the aircraft carrier USS *Wasp* (CV-7). Most were curious to see the interior of an aircraft carrier and also get out of doing chores. McKinney noted that signal flags were draped around the carrier's hangar deck, which he felt ran counter to the solemnity of the occasion. He also was impressed with the chaplain's "rip-roaring" message.

Quite a few sailors stayed ashore to take advantage of schools offered at the naval base there. Several of the signalmen, for example, attended a sound school where they learned to operate the "QC" underwater sound detection equipment. The men rotated through three rooms: one simulated a surface ship pilothouse, one simulated the control station on a submarine, and the other room tracked the simulated movements of the surface ship and sub as they tried to track each other using sound. Once the men became comfortable with the gear, they returned to *Atlanta* to use the installed gear to track an "S"-class submarine that made runs at the cruiser.

A member of the band recalled one particularly horrific night on the town. He managed to catch the only afternoon liberty boat to leave the ship before the fog moved in. After an hour ride, the coxswain pulled the boat up to the fleet landing and discharged this fellow and other dampened shipmates. Once ashore, it started to drizzle, then pour. Uncomfortably wet, the sailors returned to the fleet landing and stood under a shelter and waited for a boat from *Atlanta* to pick them up. A few boats from other ships came by but not from the cruiser. At 0200, Jim Sutherland led a group of his shipmates to the Portland Armory where a cot and blanket could be had for fifty cents. The next morning they returned to Fleet Landing only to find a large crowd of stranded sailors from several ships milling about. The weather had worsened and no boats were operating. Finally, at 0800, an ocean tug appeared, took off the sailors, and proceeded to go from ship to ship, dropping off sailors. The *Atlanta* crewmen, stuck standing on the deck of the tug, spent an hour and a half in drenching rain, waiting to get back to their ship which was, unfortunately, anchored the furthest away. "Our caps, peacoats, pants, shoes, and socks were nothing but running water … We of the returning liberty party did little that morning but get thawed and dried out."[50]

Clearing Casco Bay on Friday March 13, *Atlanta* conducted some gunnery and engineering "crashback" drills to challenge the Black Gang—a term that hung on from the days of coal—on how fast they could get the ship's screws to reverse their spin. The light cruiser made a short call to the Boston Navy Yard and then ventured back out into Massachusetts Bay to take a southerly heading to the Cape Cod Canal, passing under the raised railroad lift bridge at 1838. After clearing the canal, the cruiser headed into the Long Island Sound for a slow night passage. With daylight, Jenkins steered his ship under the Bronx-Whitestone Bridge, through Hell's Gate, and down the East River to return to pier G berth 13 at the Brooklyn Navy Yard. Perhaps many who looked across the way may have thought they were staring into a mirror as sister ship *Juneau* was also in port.[51]

Navy Yard workers swarmed aboard to fix additional deficiencies unveiled during the shakedown and make needed corrections. For example, mess deck space was converted into additional berthing to accommodate more crewmembers. Life rafts were strapped to the sides of gun mounts and other exterior bulkheads. Workers added scuppers on the bridge wings to provide for better drainage to prevent lookouts from having to stand in pools of water during and following rainstorms. While up on the bridge the workers lowered the flag bags and made access to them easier.

K-Guns, capable of firing 300-pound depth charges abeam, were placed on the stern to augment two racks containing 600-pound depth charge barrels designed to roll off the fantail. Also back aft, workers installed an additional 1.1-inch gun mount to protect the ship from enemy aircraft sneaking in from directly astern.[52]

As the ship underwent alterations, the "dope" on *Atlanta*'s future assignments remained speculative. Many crewmembers assumed the distribution of cold-weather gear meant the cruiser would fight in the North Atlantic against the Nazi submarines and aircraft. Many of the wives shared this belief. Betts Perkins's sources informed her that *Atlanta*'s homeport would be Norfolk. The ship's senior leaders knew otherwise.[53]

For those who grew up in the New York area, the period at the Brooklyn Navy Yard provided precious final moments with family and friends. Betts Perkins, who was well into a pregnancy, had to be hospitalized in Boston on her return trip from Portland to prevent a miscarriage. The XO allowed her husband to rush up to the hospital to be at her side. She survived the episode to return to New York on the eve of *Atlanta*'s final departure.

Early on March 18, a large delegation of naval and civilian inspectors from the Board of Inspection and Survey (InSurv) arrived to ride *Atlanta* for what would be her final acceptance trial. Led by the president of the board, Rear Adm. David M. LeBreton, his team of some four dozen individuals spread out throughout the interior of the ship as *Atlanta* got underway for a speed run to determine the ship's maximum speed and other characteristics, now that she was fully loaded. One enlisted observer noted, "There was gold braid all over the ship, even admirals in dungarees, climbing in and out of machinery spaces." As the ship cleared the narrows, Blimp K-6 soared overhead to provide additional anti-submarine cover. Despite rough seas and many upset stomachs, this fellow reported that *Atlanta* attained a speed of 34 knots. Upon passing back through the narrows, the light cruiser stopped off Tompkinsville off the northeast tip of Staten Island to let off the inspection party who were boated ashore and then proceeded back to Navy Yard New York and pier G.[54]

Understanding what lay ahead, Captain Jenkins generously granted leave and liberty to allow crewmembers to spend time with families. Yet, work continued to ready the ship for sea and from March 24 through 26, the light cruiser once again found itself high and dry in dry dock #4. Apparently, the InSurv team confirmed the observations that were made during the earlier full-speed runs. Bob Graff remembered, "We had triple-bladed propellers which were huge, and they sent currents of water around the ship that just knocked it ass over teakettle." During the drydocking, the pair of three-bladed monstrosities came off and yard workers installed two smaller four-bladed variants. While the smaller screws meant *Atlanta*'s top speed would be 30 knots, the loss of high-speed vibration justified the switch.[55]

Returned to pier G, the light cruiser refueled and departed Brooklyn for the last time the next morning on March 29. The cruiser would remain in New York waters at Gravesend Bay for a few more days to load additional ammo and run a degaussing range. For many of the crew's female companions, seeing the ship lie off Coney Island for three days just added to the sadness.[56]

CHAPTER 3

En Route to Hawaii

On April 3, a tug came along to off-load some of the senior officials who had observed *Atlanta*'s final preparations prior to her deployment. Along with the observers, members of the band and their instruments came off the ship. Embarking on the ship were a good number of raw recruits, who would have to be trained to perform various duties and others who were being taxied to the Canal Zone for further assignments there. Fortunately, the long cruise to Hawaii would provide ample opportunities to train.

With the anchor raised, Captain Jenkins took the precaution of sending his crew to General Quarters as the ship passed the Ambrose Lightship and came about onto a southeasterly heading. Once out into the Atlantic, Jenkins announced to the crew using the 1MC general announcing system that the immediate destination was the Panama Canal. The skipper also promised to keep the crew informed of upcoming operations upon every port departure. In addition, he reminded *Atlanta*'s bluejackets of the continuing German submarine threat, so they needed to keep alert.[1]

Captain Jenkins had a good-size audience steaming with a full wartime complement of 670. They were led by a wardroom of nearly three dozen officers, divided between regular and reserve officers. The Chiefs' Mess, commonly referred to as the "Goat Locker," carried a similar complement of senior enlisted.

Atlanta made rapid progress down the East Coast despite the implementation of a zigzag scheme. As before, Captain Jenkins sent his crew to their General Quarters stations at dawn and at sunset when lighting conditions silhouetted the ship on the horizon. On Easter Sunday April 5, with the crew enjoying outside temperatures that were in the 70s, Palm Beach was sighted to starboard at noontime. Later that afternoon, while off Miami Beach, the lookouts could spot bathers lying out on the sand. Passing Key West light to starboard at 2220, the light cruiser continued to steam along the north coast of Cuba past Havana, rounding the western tip of that island nation to head south to the canal. Along the way, one of the lookouts believed he spotted a submarine. The crew immediately responded to the General Quarters alarm. A K-Gun was fired, throwing a depth charge 50 yards to starboard. The explosion shot up a huge column of water. Not forewarned in the after-engineering compartment, Fireman Charles Dodd recalled the considerable consternation: "We the engineers, thought we had been torpedoed and all made a mad dash for the single ladder."[2]

Suspecting that the approaches to the canal on the morning of April 8 could be U-boat infested, Captain Jenkins again sent the crew to their battle stations. While the precautious step was proper, Ensign Corboy observed it probably was not necessary. He later wrote: "As we approached the canal, we were thankful not to be an enemy craft, trying to slip in. For days before we arrived aircraft were almost constantly on the horizon. Our surface ships were out in force."[3]

Atlanta's anchorage in Colon Bay at 1004 on the 8th generated some celebration within "E" Division. Bill McKinney had 04 in the anchor pool and his luck led to a 50-dollar payoff. Topside, *Atlanta*'s crewmen looked out at the gun emplacements and barrage balloons along the shoreline.[4]

★★★

For the United States, the existence of the Panama Canal proved critical to the war effort as ships were rushed from East Coast shipyards to the waters of the Pacific. Proposals for a marine crossing of the isthmus had been envisioned in the 19th Century and the French initiated an effort that fell victim to underfinancing. The national security implications of a lack of crossing became painfully apparent during the Spanish-American

War when the battleship *Oregon* steamed from the West Coast around the tip of South America to reinforce the U.S. Navy's blockade of Cuba against the Spanish Navy.

However, while the completion of the canal enhanced the mobility of the Navy, it also placed design constraints on the larger ships. For example, the beam of battleships and aircraft carriers was limited to 108 feet. At that width, the canal provided for a foot of clearance on each side.

For *Atlanta*, clearance wasn't a problem. The pilot, Captain F. A. Dear, of the Canal's Marine Division, embarked and joined Captain Jenkins on the bridge and offered guidance as the cruiser approached the entrance of the canal. Army soldiers stationed in the Canal Zone gawked at the sleek new ship that coasted past as *Atlanta*'s sailors looked back. Next, the light cruiser passed through the Gatun Lock complex in three steps for a total lift of 85 feet.

The passage through freshwater Gatun Lake offered an opportunity to flush out the ship's condensers, firemains, and toilet flushing systems. Having crossed the lake, *Atlanta* passed into the Gaillard Cut, an 8-mile man-made canyon at the Continental Divide and then the ship descended towards the Pacific via the Pedro Miguel Lock, followed by a pair of locks at the Miraflores complex. It was there at approximately 1814, that a sudden vibration on the port shaft could be felt following clearance of the Miraflores locks. The Black Gang immediately secured the port engine and *Atlanta* pushed ahead just using the starboard shaft with the helmsman adjusting rudders to compensate for the drag. Arriving at Balboa, *Atlanta* took on 224,764 gallons of black oil and on the following morning a diver inspected the port propeller and told Captain Jenkins that one blade "was bent back at a distance of 4 feet from hub cap for a length of about 8 inches." *Atlanta* thus would spend a little time in the nearby Navy dry dock to get the dent banged out.[5]

For the crew the unexpected delay meant an unplanned liberty call. A port and starboard in-port watch bill enabled half the crew to go ashore. At that time Balboa, and nearby Panama City, had a reputation among the world's seafarers as places that could cater to a wide assortment of vices. Walking into any bar, sailors often found several young ladies beckoning them with smiles.

For those who wanted to do more than drink, the Navy and Army established a prophylactic station and issued orders to soldiers and sailors intending to visit the brothels to check in prior. The Coconut Grove strip was a wide-open red-light district that featured a main drag lined by one-story garage-type huts. The brightly lit interior of each hut shined out to the street via open doors. Inside the doors, a woman sat on a rocking chair adjacent to a bed, calling out to potential customers passing by on the street. If interest was expressed, the gal would negotiate costs associated with the various services and with agreement, the man was invited in, and the doors were closed.[6]

Bill McKinney, decked out in his summer white "crackerjack" jumper, counted himself with the first batch to cross the quarterdeck that first night and head out into Balboa. Making a beeline to the first bar in sight, he muscled up to the bar and demanded the best rum and received a retort from the bartender that there was no bad rum! He then made his way to the Coconut Grove district and came upon the one-room brothels just described. Coming upon a hut with no occupant, McKinney leaped up and sat on the rocker and bantered with other sailors walking by. The Shore Patrol was not amused and told him to move on. For *Atlanta*'s sailors, liberty expired in Panama City at 2300 and secured at Balboa at 2330. Invariably, some of the crew broke curfew and found themselves in the local jail.[7]

That *Atlanta* could undergo repairs must be credited to the completion of the largest dry dock the Navy had ever built up to the time, a mere two years after the opening of the canal. On Saturday April 11, *Atlanta* eased into the 1,076-foot-long graving dock and was positioned over blocks that had been placed to conform with the hull. With the caisson placed back across the entrance of the dry dock, pumps started sucking the water from within and the cruiser slowly sank until she rested on the blocks. As the water receded, vast areas of the underside where the paint coating had fallen away were exposed. However, a new paint job would have to wait for another time.[8]

Once again high and dry, *Atlanta* seemed like a hospital patient on life support systems. Tubes and cables carried steam, fresh water, and electricity from ashore. The heads were secured as there was no salt water circulating to carry the waste over the side, nor was waste going overboard desirable with workers below. An inconvenience during the day, this arrangement became extremely burdensome to those drunken crewmen having a nature call in the middle of the night. One intoxicated sailor fell off the gangway, partway into the dry dock. Shipmates attributed his survival due to his high inebriation. Having suffered a cracked skull, this misfortunate bluejacket would have to be left behind. For others who misbehaved ashore, they would appear before Captain Jenkins at Captain's Mast. In some cases, Jenkins opted to convene a summary courts-martial.[9]

With the repair made and *Atlanta* refloated, burners were relit to start the process of building the needed steam to turn the ship's turbines. Departing for Pearl Harbor on the Sunday morning of April 12, the crew once again answered the call to General Quarters.[10]

Clearing Panamanian waters, *Atlanta* came to a base course of 280 degrees. Captain Jenkins once again addressed the crew to announce this was a heading that would take them to Pearl Harbor. While some were disappointed that the ship would not be pulling West Coast liberty, all were happy to find the Pacific Ocean allowed for a smooth ride, especially those who were still hungover.

En route to Hawaii, *Atlanta* received orders to steam past Clipperton Island, located in the eastern Pacific Ocean approximately 1,500 miles west of Nicaragua. The island was a low-lying atoll surrounding a freshwater lagoon that was once claimed for the United States by the American Guano Mining Company in the 1850s. However, the atoll had been claimed a century earlier by the French, and the United States had no interest in challenging French sovereignty. Uninhabited, the island could potentially serve as a clandestine Japanese refueling station for submarines targeting shipping emerging from Panama Canal. Captain Jenkins and his crew would check on the presence of any nefarious activities.

Atlanta's skipper chose to approach the island at daybreak when it would be difficult to spot the approaching cruiser on the eastern horizon. Already called to General Quarters as part of the daily predawn routine, *Atlanta*'s gunners stood ready to unleash a barrage on an unsuspecting submarine or merchantman. As the morning light bathed the atoll, *Atlanta*'s lookouts caught a glimpse of a vessel. Continuing to close, Captain Jenkins observed the mysterious vessel to be a white-painted schooner with unrigged sails. A boarding party was mustered, led by Lieutenants Perkins and Smith. With submachine guns and other weapons issued, the two officers and several senior petty officers were lowered over the side in a motor whaleboat.

As the small boat came alongside the suspicious schooner, Lieutenant Perkins and some of his armed enlisted shipmates climbed aboard to find the cook and a deckhand. Through an interrogation, Perkins learned that his boarding team had embarked on the *Skidbladnir*—a shark fishing schooner and that the schooner's captain, first mate, and another crewmember had taken a small boat ashore to explore the remnants of the guano mines the previous day. Unfortunately, gusts caused *Skidbladnir* to drag her anchor off Clipperton's narrow shoal and now the schooner drifted helplessly in deep water as the two men aboard lacked the skill to set sail. Those same gusts kicked up the surf on the atoll, temporarily stranding the three other shark hunters on the beach.

With his New England maritime background, Perkins instructed the boarding party sailors on how to sail a schooner. With the free-hanging anchor hauled in, Perkins sailed *Skidbladnir* close enough to the atoll to again drop the anchor to get a firm hold. With this good deed done, Perkins wished the cook well, and the boarding party returned to the light cruiser. With Clipperton cleared of any enemy activities, Jenkins again turned his warship on a westerly heading.[11]

From Clipperton Island, *Atlanta* traversed thousands of miles of open ocean at a speed of 15 knots, to conserve fuel. At such speed the trip took approximately a week's time. Every third day the quartermasters moved the clocks back an hour to conform with the changing time zones. Steaming alone in potentially sub-infested waters, Captain Jenkins took no chances. The cruiser zigzagged back and forth along the base course. Every morning the crew awoke before twilight to the sound of General Quarters. At sunset the crew again scrambled to their battle stations and remained in place until darkness hid the cruiser from underwater attackers.

One afternoon, the crew went to General Quarters and ran through a half-hour battle problem. At each station, an observer held a stopwatch and slips of paper. For example, at a minute and 35 seconds into the drill, a slip reading "shell through pilot house, helmsman killed" was handed to the helmsman. He fell away from the wheel and the bridge team quickly found someone to replace him. Meanwhile, firefighting crews entered the space to simulate dousing the flames and clear out smoke. Bob Graff saw himself and the other junior officers as coaches in what, upon reflection, were team-building exercises. "You had a job to do and when the shells started flying and the bombs started dropping you still had a job to do … and the first time when *Atlanta* went to General Quarters in the Pacific, it just functioned smoothly, and we hit airplanes and brought them down."[12]

In addition, the crew had daily gunnery practice with the 5-inch 38 guns and the smaller arms. As in the Chesapeake Bay, a balloon would be released and once it had risen a mile away, the gun batteries unleashed salvos. However, the gun crews were now beginning to have success with knocking them down.

Along the way, *Atlanta* picked up a feathery escort. One day an albatross gracefully glided alongside. The dark long-billed bird occasionally turned to exchange glances with the topside sailors. Then towards sunset, this winged companion turned to the south and bade farewell.

Atlanta's track took her to the west of the Big Island of Hawaii before she turned up to Oahu. Consequently, crewmen having exterior battle stations at sunrise on Saturday the 25th were treated to the spectacle of the sun coming up over Mauna Loa. At that time, the destroyer *Maury* appeared on the horizon as an escort. *Maury*'s appearance pleased many who worried that *Atlanta* could be attacked by friendly aircraft patrolled from Pearl, whose pilots would have never seen a ship of such design. Graff recalled a pilot boat came way out to greet them and then they had to wait as other ships were entering and leaving the channel.[13]

Once *Atlanta* passed through the minefields guarding the entrance of the channel, Ed Corboy noted that, "all hands strained their eyes for a sight of what the Japs had done to us in their sneak air raid of Dec. 7." The blackened hangars at Hickam Airfield offered the first evidence of the attack. "Then came the litter of masts and turrets on the beach to our port side." Upon seeing the hulks of the destroyers *Cassin* and *Downes*, Corboy observed: "Their steel sides, ripped by the infernos that had gutted them, seemed to cry to us for vengeance."

As the cruiser glided forth past oil-soaked shorelines and came upon the wreckage of Battleship Row, Corboy wrote, "The USS *Oklahoma* lay on her side. At her stern was the *West Virginia* still sitting on the bottom, but ready for raising."[14] The reality hit Graff from his perch on the signal bridge: "Boy, all of the sudden we were in the war ... it was the most painful sight I have ever seen in my life ... I remember everybody aboard *Atlanta* was in tears."[15] One of Graff's signalmen described *Arizona* as "terribly burned and blackened, with her mainmast at a precarious angle."[16] Still, the damaged ships flew their stars and stripes which left an impactful impression on all.

Since December 7, much preliminary work had been accomplished to return most of the damaged warships back to service. Neither *Arizona* nor *Utah* was salvageable. *Oklahoma* would also not return to service. Thanks to an enormous series of pulley rigs erected on Ford Island, the capsized hull would be flipped over. Once refloated, the Navy determined the old battleship's only value to the nation would be her steel sold for scrap. However, while being towed to the West Coast to be broken up, a storm swamped the carcass of the old BB, and the remnants remain on the bottom of the Pacific today.

Meanwhile, *California* had entered dry dock #2 on April 9, having suffered damage caused by two torpedoes, a direct bomb hit, and some near misses. For BM1 Spadone looking over at the one remaining cage mast that could be seen rising from within that dry dock, his thoughts must have been with those former shipmates who were killed and wounded during the Japanese attack.[17]

The cruiser continued into the East Loch to her assigned mooring C-6 alongside the cruiser *Nashville*. The *Brooklyn*-class cruiser had just returned as part of Doolittle mission centered around *Hornet* and *Enterprise*. *Atlanta*'s crew could not help but note that paint had blistered off several of *Nashville*'s 6-inch guns which had been employed to sink a Japanese picket vessel which had discovered the approaching American task force.

The destroyer tender *Dixie* floated in an adjacent berth on the starboard side and the repair vessel *Vestal* could be seen off the port side. On board the light cruiser clocks were adjusted 30 minutes to conform with Hawaii time, which was 9 and a half hours earlier than Greenwich Mean Time. Having arrived in paradise, *Atlanta*'s sailors looked forward to liberty ashore.[18]

CHAPTER 4

Pearl Harbor Out and Back

For *Enterprise* and *Hornet*, and the other ships assigned to Vice Adm. William F. Halsey's Task Force 16, there would be little time to celebrate the raid by the twin-engine Army Air Forces B-25 Mitchell medium bombers against the Japanese home islands. Adm. Chester W. Nimitz, acting on a prediction from Cdr. Joseph Rochefort that the Japanese Fleet intended to move into the Coral Sea in early May, ordered Rear Adm. Frank Jack Fletcher to rendezvous with *Yorktown* and *Lexington* in the Coral Sea to prepare to thwart what would be a Japanese plan to land troops at Port Moresby on the southern coast of New Guinea. Though thousands of miles had to be traversed before Halsey's carriers could arrive to augment Fletcher's, in an extended multi-day sea battle, Task Force 16 had the potential as serving as the cavalry arriving over the horizon.[1]

Atlanta had been rushed to the Pacific Fleet with the hope that she could steam with Halsey's task force. However, Captain Jenkins had to confront some deficiencies that would sideline his ship in the short term. During practice firings en route, the gunner's mates discovered that the hydraulic gear for the 5-inch fuze-setting shell hoists would burn—not an acceptable situation moving live ammunition within a confined space. With no drawings available and no local expert who understood how the gear was supposed to function, Lieutenant Commander Nickelson awaited the arrival of a "Mr. W. Cody of the Northern Pump Company" of Grantsville, Wisconsin.[2] In addition, Lieutenant Commander Loeser reported the deterioration of tubes carrying cooler ocean water through his condensers which served to take the expended steam that just passed through the ship's turbines and condense the vapor back to water to feed again into the boiler. Should salt water leach into the engineering plant's steam cycle, the consequences would be catastrophic with corroded tubes failing in the boilers. This was a challenge the Pearl Harbor Navy Yard could handle.[3]

Captain Jenkins must have found the mechanical and engineering problems frustrating. However, while many of the crew were itching to get after the Japanese after viewing the devastation on their inbound trek into the harbor, the prospect of extra liberty in Oahu likely caused little heartburn. Unfortunately, for the third of the crew that was allowed to go ashore on a given day, they would find the liberty situation somewhat disappointing. First, to enable Honolulu to maintain a blackout at night, the Navy only allowed sailors on the streets during daylight hours. Liberty hours in Honolulu lasted from 0900 to 1800 with the bars closing at 1600.

In addition, *Atlanta*'s sailors found they were not alone on the streets. Since the December attack, the number of military personnel on the island had tripled. Tent camps appeared all over the island. The vibrant downtown hotel district associated with the 50th state was still decades away. Honolulu of the Depression-era 1930s hosted a mere half dozen quality hotels and was a secluded vacation destination, lacking the infrastructure to support the entertainment needs of thousands of mostly young males from the mainland.

Many of the men rushed to the phone exchange to place calls to loved ones back in the States. Strict censorship applied to the callers who were warned that operators monitoring the lines would cut off calls with the slightest violation. No names, dates, places could be mentioned. Betts Perkins was exhilarated to hear from Van Ostrand. However, after the initial greetings, Betts heard the line go dead after her husband had blurted out "Oh Rabbit Kilud." Jane Shaw experienced the same abrupt cut off when she attempted to tell her husband she was heading to Fort Benjamin Harrison.[4]

For those sailors not desiring to phone home, Hotel Street became a popular destination. Along the one-mile strip, an assortment of entertainment establishments competed for a piece of *Atlanta*'s payroll. One of the larger centers drew the bluejackets in with a swing band, shooting galleries, pinball machines, and a soda counter. Other joints lured young men from the mainland by offering to pose them for pictures with hula shirt-clad natives or offering souvenirs that often were produced from back home. Beer and hard liquor flowed from behind the numerous bars that lined the strip. Nearby the bars, tattoo shops etched anchors, American Eagles, images of pretty girls, and other designs into the skins of hundreds of customers a day.

For those seeking more sinful pleasures, there were some 20 registered brothels with 250 women to cater to sexual needs, and numerous unregistered houses. A section of Honolulu bounded by Hotel, River, Nuuanu, and Kukui Streets hosted 15 of these residences. Given the influx of young men seeking an opportunity to be with a woman before going off to war, business was brisk. One of *Atlanta*'s sailors wrote: "Long lines of expectant sailors formed every morning outside of several establishments not recommended by the chaplains. These sailors stood in line reading the morning paper, getting shoeshines from urchins, and generally behaving as men waiting for a bus. I learned that business inside was conducted with assembly line speed. Each customer was allowed three minutes, and those unable to consummate the deal in the time allotted were invited to come and try again."[5]

Not overly concerned about the morality of its men, the Navy worried more about the spread of venereal disease. On *Atlanta*, division officers told their men that if they intended to indulge in shoreside pleasures, they had to report to a pharmacist's mate to receive prophylaxis. Those who refused to check in with the "Doc" and subsequently became infected could expect to be punished and lose pay. Remarkably, *Atlanta*'s sailors complied as during this and later liberty periods, only one man of the cruiser's near-700 crew came down with an infection.[6]

Rather than heading to the sailor- and soldier-crowded bars and restaurants, Perkins, Shaw, and Pat McEntee decided to hit the beach. To their surprise, they found themselves working through several mazes of barbed wire and other obstructions to reach the crystal blue waters.[7]

For the two-thirds of the crew staying aboard, in addition to resolving the ship's mechanical and engineering problems, they worked to boresight the guns, bringing sights and guns into perfect alignment. For some crewmembers, likely thanks to contacts made by Petty Officer Spadone, tours were arranged to visit *California* as she underwent repair in dry dock #2 to enable her to get seaworthy again.

On the morning of May 6, with repairs made, *Atlanta* departed for waters off Maui and Kahoolawe to determine whether the adjustments made improved the accuracy of the guns. The 5-inch, 20mm, and 1.1-inch guns crews each took their shots at sleeve targets as they were flown by. For a final test, *Atlanta*'s 5-inch 38 batteries successfully knocked a radio-controlled drone out of the sky. The demonstrated marksmanship enhanced the crew's confidence as reflected in the referrals to the ship as the "*Mighty A*."[8]

Back at berth C-6 on the following morning, *Atlanta*'s gunners were led to believe that those newly honed skills would be put to the test as a lookout on *Dixie* spotted unidentified aircraft and warships throughout the basin, and sounded the general alarm. As hundreds of eyes peered into Oahu sky, Captain Jenkins's ship prepared for getting underway. Those preparations were called off shortly after the unidentified planes were deemed friendly.[9]

After three more days at Pearl Harbor, *Atlanta* left as part of Task Group 16.19. On the night prior to departure, a steward picked up a stray black-and-white terrier at Fleet Landing and snuck the pooch on board. As the ship headed southward, many of the crew were startled to see the dog running up on the forecastle. Noting the positive effect on morale, Captain Jenkins dropped any reservations he may have had about the animal. Named "Lucky," the ship's new mascot had a lifejacket made for him and a battle station assigned. Lucky also received an ID card featuring a paw print and photo.

Soon a ghosted autobiography appeared in *Chock "A" Block*, the ship's newspaper that claimed the dog's father was a Marine Corps mascot and the mother was of German roots. Ensign Corboy observed: "The appearance of Lucky was the last circumstance necessary to weld the *Atlanta's* company—officers and men—into a solidly united, fighting family. In this respect alone, Lucky was worth his weight in gold."[10]

Task Group 16.19 consisted of the destroyer *McCall*, which led the way out, followed by *Atlanta*. In their wake came the ammunition ship *Rainier* and the oiler *Kaskaskia*. Once on the high seas, Captain Jenkins announced on the 1MC that they were escorting the two ships to Nouméa, New Caledonia, where *Rainier* and *Kaskaskia* would off-load their cargoes. Within the four-ship formation, Capt. William W. Meek in *Rainier* had graduated the Naval Academy two years ahead of Jenkins and with that seniority served as the officer in tactical command (OTC). Meek positioned *Atlanta* and *McCall* ahead off the bow of his ship to starboard and port. The oiler followed in his wake. This Y formation enabled the two combatants to use their sonar to best effect.

With the four ships' projected track placing them within range of Japanese airfields in the Gilbert and Marshall Islands, tow planes carrying target sleeves flew out during that first day to provide additional target practice. *Atlanta* knocked down one tow plane's sleeve on the first pass. *McCall*, *Rainier*, and *Kaskaskia* gun crews banged away at the other sleeves.[11]

If there were any jitters about the chances of coming under air attack, the smooth seas had a calming effect. The constant rainbow in *Atlanta*'s bow spray was only occasionally faded when the cruiser passed near or through the many roving rain squalls. All around, silver flying fish skirted across the wavetops. Some of the junior officers also worked to break the tension. Lt. (jgs) Ira Wilson, Jack Pierce, Dusty Rhodes, and Ens. Bob Graff organized boxing lessons, sunbathing sessions, and group sing-alongs for the enlisted sailors. The bluejackets enjoyed crooning such tunes as "Sam, You Made the Pants Too Long."[12]

Three days out, reports of a submarine contact sent the crew running to their General Quarters stations. While sonar operators on *McCall* and *Atlanta* attempted to confirm the contact, a lookout reported sighting an aircraft. After both the submarine and airplane contacts appeared as false alarms, eyes turned towards eastern skies where an airborne shadower could clearly be discerned. With additional false submarine contacts, most of the crew was beginning to get a bit edgy for two reasons—first, the threat posed to their personal life longevity due to enemy action, and second, the immediate threat to bodily harm pending as *Atlanta* approached the equator.[13] Electrician's Mate McKinney, who had attended college prior to joining the service, recalled: "... I learned that the transition from pollywog to shellback is an exercise that makes a college fraternity initiation quite tame by comparison." At mid-afternoon on the 13th, one of the junior "pollywog" officers found himself standing atop of the third 5-inch 38 gun mount, just forward of the bridge, wearing a winter peacoat and shorts. He was provided a pair of brass fire nozzles, that had been affixed together in the shape of binoculars, to scan the horizons for Davy Jones. Another crewman, also donning winter attire and carrying a roll of toilet paper, relayed reports from the lookout down to the engine room.[14] Late in the afternoon, Davy Jones did arrive, crawling up the hawsepipe.

May 14 started auspiciously early with a 0330 rush to General Quarters as an aircraft could be heard overhead. The crew hardly had a chance to return to their bunks when the daily dawn GQ was sounded at 0515. Seventy-five minutes later, another suspected sonar contact would prove false.[15] Yet that combination of hostile threats would not defer a long-cherished nautical tradition—the crossing of line ceremony. Following morning muster, all hands donned pirate outfits, many pulling cloth scraps from the engineering department ragbags to create their costumes.

The crew was called to quarters aft on the main deck and a pirate flag was hoisted on the mainmast. There they stood at attention in dreadful anticipation of the arrival of King Neptune and his strangely-dressed retinue of loyal shellbacks. At 0900, Neptunus Rex and his royal entourage arrived. With the pollywogs lined up for a personnel inspection, each would be doused with a greasy graphite concoction on the face and hair by one of King Neptune's minions while another member of the party gave impromptu haircuts.

The shellbacks set up a formal receiving line on the starboard side, adjacent to furthest aft 5-inch 38 gun mount. The pollywogs stripped down to their waists and lined up barefoot to await their fate that first called for running through a gauntlet of shellbacks flailing rope or canvas-covered water-soaked truncheons. Upon surviving the gauntlet, each pollywog confronted his highness himself. To add to the agony, water from the fire main moistened the deck to enhance the effects of live wires that were thrust towards the sorry victim as King Neptune determined the ultimate punishment he should suffer. Depending on the King's decision, the unfortunate pollywog could have been sent over to one or more stations that had been set up by the depth charge racks. Shellback barbers provided additional "bad" haircuts, cutting clumps of hair, and many of the pollywogs had to undergo "an operation" lying on a surgical table while a "doctor" applied an electrified scalpel to various parts of the victim's anatomy.

Finally, the pollywogs had to crawl through a 30-foot-long canvas chute that was 3-foot wide in diameter. Before entering the chute, each pollywog endured another a shock from King Neptune's electrician. Bob Graff recalled:

> Pollywogs were forced to crawl through the tube and as they crawled through, they were whacked repeatedly on the butt, and on their back, but not their heads, by shellbacks with homemade truncheons. Most of the shellbacks were genial enough but every so often there was a tough guy who wanted to prove he was somebody and really whacked us.[16]

It did not end there. The pollywogs now had to drink a vile concoction and receive additional verbal abuse before finally—King Neptune ended the ceremony and declared all of the former pollywogs to be shellbacks. Each new shellback received a graduation document that would be protected as if it were a birth certificate as insurance against ever having to endure such a ritual again.

The XO and the ship's Doc observed the event to make sure that the shellbacks were not overly abusive. Most of *Atlanta*'s crew had to endure sore buttocks for the next few days and barbers back in Honolulu later would be challenged to redo many of the altered hairstyles. Unfortunately, there were one or two sailors who suffered severe strains or broken bones due to the festivities.

Given the number of pollywogs and the frequent interruptions due to General Quarters, the initiation into the realm of the deep lasted until 1400. Much of the remainder of the day was dedicated to cleanup.

The first full day in the Southern Hemisphere, an aircraft sighting again had *Atlanta*'s sailors scurrying for their GQ stations. This time a friendly Catalina PBY patrol plane proved to be the culprit. *McCall* had another alleged submarine contact that again sent sailors scrambling, and then there was another aircraft sighting that proved to be a seagull. Shortly thereafter, Lt. Jack Broughton held additional lookout training. Finally, the four-ship formation came into contact with other American naval ships passing in the opposite direction: the destroyer *Gwin* followed by the near-empty oiler *Sabine*.[17]

That evening, Captain Jenkins learned that *Atlanta* had received new orders. On May 16, following a morning refueling of 140,000 gallons from *Kaskaskia*, *Atlanta* veered on a more westerly heading to join up with fleet units returning to Pearl Harbor. With the crossing of the International Date Line, May 18th did not occur. So after two days of independent steaming, early on May 19, with the crew at their early morning precautionary GQ stations, those having topside battle stations were treated to a spectacle.[18] Reinforced by Coral Sea combat veterans, the sight of Task Force 16 coming over the horizon awed topside sailors standing at their battle stations. Seaman 2nd class Walter Hollingsworth, perched in the crow's nest on the mainmast high above, caught the first glimpse of an oncoming carrier off the starboard bow. Minutes later peering out from his station in the forward gun director, Ensign Corboy caught sight of the advancing task force. He wrote, "It was the greatest thrill of my naval career up until then." Corboy waxed on:

> We in the director now could see the carrier, looming big and black. Then the masts of the cruisers came up, along with another carrier and finally the thicket of destroyer masts. Now the hulls began to show, with the great carriers dominating the scene. It was the grimmest, blackest armada I'd ever seen. There were ships all the way across our bows ... Up to that grim bunch of battle bulldogs, waltzed the Atlanta, as beautiful and graceful as a deer.[19]

Perhaps the sailors of that floating armada would have been more pleased to see the Navy's battleships come over the horizon, however, the sleek new light cruiser would do just fine as within its hull lay some 80 bags of mail! Two of the screening destroyers took turns alongside to receive the sacks of news from home. With adroit shiphandling skills, the conning officers steamed their tin cans close abeam to allow the canvas sacks to be hauled over attached to a manila line. With Coral Sea veteran *Phelps* alongside, there was a vigorous exchange of shouts as *Atlanta* got the dope on the recent engagement that cost the United States the loss of *Lexington* as well as the oiler *Neosho*.[20]

Assigned to Cruiser Division Five that consisted of the *Northampton* and *Pensacola*, *Atlanta* was given a station off the starboard bow of carriers *Enterprise* and *Hornet*, which rendered the ship's antiaircraft capability ineffective against the anticipated direction of a potential enemy air attack. After two days (with May 20 occurring twice thanks again to crossing the International Date Line) the OTC, Rear Adm. Raymond Spruance in *Northampton*, became enlightened and moved *Atlanta* and her AA batteries to the front of the formation.[21]

Over the next few days *Atlanta*'s watch standers became acquainted with operating carriers as on the 22nd, Task Force 16 passed back into the Northern Hemisphere. No ceremonies at this juncture.

On both May 23 and 25, those topside witnessed an air show as torpedo planes off *Enterprise* launched a simulated attack against the formation. The torpedo plane runs benefited both the aviators and the gun crews on board ships who gained experience in tracking and slewing the AA batteries onto the approaching aircraft. As the planes buzzed overhead, Captain Jenkins learned of naval intelligence warning of a pending Japanese move against the Aleutians and Midway Island.[22]

To enable the various ships to get some live-fire exercises in, Halsey strung his ships into a long line with some of the cruisers in the lead, followed by some of the destroyers, then the carriers and oilers, followed by more destroyers and cruisers. Without warning, the carrier guns fired off at different heights and vectors. When the shell burst into a black puff, gun crews on designated ships up and down the line responded by firing shots at the residual smoke. Corboy observed: "Some of the gun crews showed up well, excellently in fact. Others were not so good."[23]

On the morning of May 26, as Task Force 16 approached the entrance of Pearl Harbor, *Pensacola* reported "man overboard." As the junior cruiser in the formation as determined by the date of rank of the commanding officer, *Atlanta* held the tail position in the column and dropped her motor whaleboat into the ocean with Lieutenant Mustin embarked to recover the misfortunate bluejacket. Sure enough, shortly after the lad was pulled out of water, a shark appeared. Unfortunately for Jenkins, the shark-infested waters were also sub-infested waters so Rear Admiral Spruance admonished him for his humanitarian action, suggesting such a rescue should have been conducted by a smaller warship.[24]

As *Atlanta* was the last of the cruisers to come into port, the light cruiser circled to the left around Ford Island and moored against dolphins immediately astern of the *Arizona* in berth F-8, the spot that had been occupied by *Nevada* during that fateful morning of the previous December. Jenkins and his Sea and Anchor detail posted up in the bridge peered down at the awash main deck of the devastated BB, the nameplate on the sternpost could easily be read.

On May 27, Captain Jenkins granted liberty to the third of the crew that had missed a turn during the previous time in. Many of those left on board were mustered to form working parties to bring loads of stores down below. Throughout the day small motorboats pulled alongside with stores wrapped in cargo nets. The ship's crane, located just forward of the aft stack, hoisted each load on board and lowered it on deck. As each net broke open, sailors swarmed like ants to remove the boxes of food, spare parts, toilet paper, and other supplies. The Supply Officer, Lt. D. C. T. Grubbs, his assistant Ens. James D. Koiner, and the senior ship's storekeepers kept track of what was coming on board and directed where the items were to be stored. Some ammunition was delivered to replenish the rounds expended during target practice. Up above, the signal gang repainted the area around bridge.[25]

As they applied gray paint to the aluminum superstructure, no doubt Ensign Graff's signalmen noted the arrival of *Yorktown*, damaged in the recent Battle of the Coral Sea. As the wounded carrier entered dry dock #1 the next day, Rear Admiral Fletcher sat down with Admiral Nimitz and Rear Admiral Spruance, now in command of Task Force 16 at the recommendation of Vice Admiral Halsey who was incapacitated due to shingles. Having been thoroughly briefed on the recent Coral Sea encounter, Nimitz met with his two task force commanders for a frank conversation of what his intelligence people were telling him and how to exploit this intelligence to inflict a blow against the enemy.[26]

Nimitz directed Spruance to depart with Task Force 16 the next day to a point northeast of Midway. Fletcher would follow in his wake once *Yorktown* was repaired and refloated.

That night *Atlanta*'s sailors peered across the horizon and could see only darkness, given the blackout conditions in effect with one exception. Across the channel, a glow could be seen from the dry dock where arc torches cut into metal on *Yorktown*. Meanwhile the crews of Task Force 16 ships got the dope on what actually happened during the first naval battle in history where opposing ships never came in sight of each other. The Japanese had scored two bomb hits and two torpedo strikes on "*Lady Lex*." However, the giant carrier had withstood these blows. It was the inability of the crew to combat two explosions caused by leaking vapors from aviation fuel tanks that proved fatal. Eventually escorting destroyers fired torpedoes to help scuttle the carrier.[27]

CHAPTER 5

The Battle of Midway

Departing Pearl Harbor on the morning of May 28, the ships of Task Force 16 reversed the order of their entry of two days earlier with the "junior" ships leading the way out to sea. *Atlanta* now led the cruisers out of Pearl Harbor. By midday, the task force had formed up and headed around the western end of Oahu before heading to the northwest. Lieutenant Shaw noted in his diary that shortly after leaving Pearl Harbor, Admiral Nimitz sent a radio message "wishing us well and expressing his confidence in our ability to strike hard and successfully at the enemy." Shortly thereafter, the signalmen on *Enterprise* started transmitting a message from Rear Admiral Spruance via flashing light that spelled out the commander's intent.

Up on the signal bridge on *Atlanta*, Ensign Graff's signalmen on watch copied down the Morse code dit-dots coming across the ways. As the signal lamp shutters fluttered open and shut, the following letters were printed on a message pad:

> AN ATTACK FOR THE PURPOSE OF CAPTURING MIDWAY IS EXPECTED. THE ATTACKING FORCE MAY BE COMPOSED OF ALL COMBATANT TYPES INCLUDING FOUR OR FIVE CARRIERS, TRANSPORTS AND TRAIN VESSELS. IF PRESENCE OF TASK FORCE 16 AND 17 REMAIN UNKNOWN TO THE ENEMY WE SHOULD BE ABLE TO MAKE FLANK ATTACKS ON ENEMY CARRIERS FROM A POSITION NORTHEAST OF MIDWAY. FURTHER OPERATIONS WILL BE BASED ON RESULTS OF ATTACKS, DAMAGE INFLICTED BY MIDWAY FORCES, AND INFORMATION OF ENEMY MOVEMENTS. THE SUCCESSFUL CONCLUSION OF THE OPERATION NOW COMMENCING WILL BE OF GREAT VALUE TO OUR COUNTRY.[1]

The new task force commander had imposed radio silence to prevent enemy radio detection finders from discovering his movements. Thanks to Nimitz's cryptologists, the Americans had a near-perfect understanding of the enemy's movements. Of immediate concern was the knowledge the Japanese intended to deploy their submarines to form a barrier between Pearl Harbor and Midway. Could Task Force 16 reach waters northeast of Midway before the barrier could be put in place? From perches within the upper decks of Captain Jenkins's ship, lookouts kept an alert eye for telltale signs of periscopes slicing through the waves.

During the transit back from the Southwest Pacific and the short turnover in Pearl Harbor, *Atlanta*'s crew learned more about the demise of one of the fleet's most beloved warships—"*The Lady Lex*." It was fire, fed by combustible materials that led to *Lexington*'s demise. *Atlanta*'s crew took the lesson learned to heart and immediately started to strip the ship of items that could burn or prove to be a missile hazard. Woodwork, curtains, and canvas awnings went over the side. Petty Officer McKinney lamented the waste of tossing 10-gallon milk cans over the side, given that they cost about $4.50 each.[2] Meanwhile, Captain Jenkins relentlessly conducted gunnery and damage control drills to hone the crew's ability to fight as a team.

Three days out Pearl, *Atlanta* maneuvered along *Cimarron* to take a gulp of 80,000 gallons of black oil. Having gained experience in underway replenishment during their recent South Pacific sojourn, First Lieutenant Sears's deck force, augmented by shipmates in other divisions, must have become adept at the hand-over-hand heaving technique used to haul in lines that would pull over a cable from the oiler. Once tensioned, a fuel line connected at intervals with pulleys would slide down the cable until the end was connected to piping leading to the light cruiser's bunkers. Hooked up steaming at a steady course and speed, the two ships were extremely vulnerable to submarine torpedo attack. Upon completing fueling *Atlanta*, *Cimarron* turned away back to Pearl, wishing Captain Jenkins and his crew "Good luck."[3]

As *Cimarron* and a sister oiler departed over the southeastern horizon escorted by three destroyers, *Atlanta*'s bridge watch scanned the horizon to see Rear Admiral Spruance's flagship *Enterprise* and sister ship *Hornet* closely screened by the cruisers *Minneapolis, New Orleans, Vincennes, Northampton,* and *Pensacola*, along with eight destroyers forming an outer screen.

Though additional reinforcements were coming as *Yorktown* and her escorts steamed from Pearl Harbor on May 30, and Midway had its own airfield, the Americans were vastly outnumbered by Adm. Isoroku Yamamoto's forces approaching the mid-Pacific atoll from the northwest. For example, the Japanese forces included seven battleships—the Americans had none. In terms of ship numbers and sheer tonnage the odds seemingly favored the Japanese, however as historian Jonathan Parshall has pointed out, with the addition of *Yorktown* and the naval air station at Midway, the Americans would have an edge in aircraft.[4]

The June 1 entry in *Atlanta*'s war diary positioned the ship at 0800 local time at 31 degrees 58' North, 174 degrees 11' West continuing on the northwesterly heading. This information was passed onto the crew by Ensign Graff. Enamored with the young officer's articulation, Captain Jenkins had his signals officer doing a five-minute morning broadcast over the general announcing system. Graff opened with "Good morning, gentlemen," and continued with an update of *Atlanta*'s position and situation as well as other news on events elsewhere on the planet that came in through the radio shack overnight. An hour past noon, the formation turned to a northerly heading of 020 degrees. As Task Force 16 arrived at a position that had been dubbed "Point Luck" some 300 miles northeast of Midway on that afternoon, that Task Force began to steer in various directions as needed to support flight operations. At 1730 the formation turned to a southeasterly heading. The diary also recorded the weather as "Overcast with intermittent fog and rain throughout the day."[5]

Arriving in the vicinity of Point Luck, Captain Jenkins took time to address the crew on the ship's general announcing system and provide an overview of what was anticipated over the coming days. Throughout the day, sailors stored lifejackets at the various battle stations and broke out ammunition. Shaw wrote:

> Everything was carefully checked. We all searched our minds wondering what our reactions would be in the heat of battle. The Wardroom and crew's quarters

> rang with conjecture as to just what was going to happen. I doubt if fear was an emotion experienced by many of the crew. Mostly it was curiosity over events to come.[6]

Throughout the evening and into the next day which featured gray cloudy skies, *Atlanta*'s helmsman received numerous orders to come right or left as Jenkins's ship maneuvered to conform with Task Force 16's frequent course changes. Back in the radio shack, radiomen operating the ship's radio detection finder began picking up Japanese voice transmissions from bearings back in the direction of Pearl Harbor. Japanese submarines, arriving at their designated barrier stations were reporting back home.[7]

Did those submarines arrive in time to intercept Fletcher's Task Force 17, consisting of *Yorktown*, the cruisers *Portland* and *Indianapolis*, and five additional destroyers? Apparently not as on the afternoon of June 2 at 1633, a lookout spotted Task Force 17 on the eastern horizon.[8] Rather than merging the two task forces together with the prospect of forcing ship commanders to learn how to operate in a framework with warships they had little operational experience with—a formula that had disastrous consequences months earlier in the Java Sea—the Americans elected to continue operating as separate tasks forces with the senior Fletcher as the overall on-scene commander.

Upon reaching Point Luck, Fletcher took overall command of the two task forces and directed Spruance to position Task Force 16 about 10 miles southward to maintain line-of-sight communications. At 1924, *Atlanta*'s log recorded Task Force 16 heading west, speed of 14 knots. Fletcher directed that the Task Force 16 carriers be ready to launch an attack on short notice. His flagship, *Yorktown*, and accompanying cruisers would have launched scout planes to search for the enemy armada.

Having been briefed on the intelligence, Captain Jenkins anticipated confirmation of Japanese movements the next day, June 3. Sure enough, reports arrived on Japanese air attacks on the American base at Dutch Harbor in the Aleutians. Armed with information that radio transmission intercepts from Japanese submarines were now coming from multiple bearings, *Atlanta*'s lookouts continued to peer for signs of a pipe sticking vertically out of the swells. With relatively calm seas, an intrusive periscope would have been easy to spot except for the appearance of numerous blown glass Japanese fishing floats that dotted the ocean surface having drifted away from the home islands. Paranoid that the Japanese had attached mines below, the light cruiser's helmsman steered clear of the glass floating balls.[9]

Early on June 3, as *Yorktown*'s scouts departed on their assigned vectors, *Atlanta* steamed on a northwest heading, as the guide ship in a formation placing the light cruiser in the middle of the task force between the two carriers. At 0920, *Atlanta*'s lookouts likely spotted the returning *Yorktown* scouts but unable to discern the aircraft markings, the ship set a precautionary General Quarters.[10] After some tense minutes, with many pairs of binoculars studying the flying objects, the aircraft were determined to be friendly. As the day wore on, the two task forces remained within sight of each other northeast of Midway.[11]

Word came that a Midway-based PBY sighted a Japanese ship formation some 700 miles west of the atoll. In response nine Army Air Forces B-17 bombers lifted off from Midway later in the day to drop bombs on the enemy ships from 8,000 to 12,000 feet.[12] Reports of the attack received by Task Force 16 and *Atlanta* must have buoyed morale. Lieutenant Shaw recorded in his diary that the Flying Fortresses claimed to have hit two battleships and score a near miss.

In reality, the B-17 bombs completely missed their targets. However, there was an audacious late-night attack made by four lumbering Catalinas that managed to place one torpedo in the hull of an oiler. Shaw would write, "... Midway-based PBYs delivered a moonlight night attack on the enemy force with considerable success"—certainly an overstatement of reality. While the initial good news may have instilled some sense of optimism among the crew of *Atlanta*, the news also meant that the promised battle was at hand. Though damaged, the enemy ship was able to stay in formation with the rest of the invasion fleet.[13]

As the American air defenders at Midway lashed out at the approaching enemy invasion forces approaching from the west, a weather front provided cover for the Japanese First Carrier Striking Force commanded by Vice Adm. Chuichi Nagumo, coming in from the northwest. The First Carrier Striking Force included *Kaga*, *Soryu*, *Hiryu*, and *Akagi*—all veterans of the strike on Pearl Harbor.

June 4—a pivotal day during the war—opened for *Atlanta* with sunrise General Quarters and as the sun came over the eastern horizon, the crew stood down to return to a Condition 3 watch bill that called for some manning of the ship's weapon systems. As those crewmembers not on watch made their way down to the chow lines, hundreds of miles to the west, the four Japanese carriers launched an airstrike aimed to eliminate the air threat from Midway's runways. Reports of the Japanese attack from Midway island filtered down throughout Task Force 16 at about 0630. *Atlanta*'s war diary recorded: "Japanese ships reported to southwest. Cleared ship for action and went to General Quarters. Maneuvered at various courses at high speed throughout the day as antiaircraft screen for HORNET while her air group made repeated attacks."[14]

To close the gap on the enemy forces, Spruance ordered his ships to head west to close the gap with the Japanese. Though Spruance's task force initially sped west, the winds did not cooperate, and the two carriers had to reverse course to a southeasterly heading to launch their aircraft. At 0700, the first plane rose off from *Hornet*. Six minutes later *Enterprise* replicated *Hornet*'s feat. Positioned near *Hornet*, *Atlanta*'s topside personnel had a ringside seat as both *Hornet* and *Enterprise* launched torpedo bombers, dive-bombers, and fighters. Sunlight bathed Spruance's warships on a day that featured exceptional visibility and a few scattered clouds. On the horizon, lookouts peered through binoculars to spot *Yorktown* recovering scout planes that had been launched before dawn. Fletcher's aircraft would follow those of *Hornet* and *Enterprise*.

Unfortunately, the clear skies enabled lookouts to spot a lone aircraft hovering to the west. A scout plane from the Japanese cruiser *Tone* had spotted Task Force 16. Appreciating that he probably lost the element of surprise, rather than wait for all aircraft to get airborne and strike in a coordinated attack, Spruance directed airborne squadrons to get on with the attack.

By mid-morning, U.S. Navy aircraft from the three carriers were en route towards a location in the North Pacific where the Japanese carrier force was expected to be. On *Atlanta*, as with every other ship within the two task forces, an intense waiting period began. A crowd of officers and enlisted sailors stood by speakers as the radio room awaited transmissions from the three air groups. Topside, lookouts kept one eye on the Wildcat fighters held back for combat air patrol duties and the other out for enemy bogeys.

Because of the report from *Tone*'s scout plane, the Japanese carriers turned to the northeast to close on Spruance, caught in a quandary, as the returning morning air strike from Midway had to be landed. In the interim, the Japanese fended off one attack after another from Midway-based airpower.

Arriving at the point where the enemy carriers were expected to be, there was nothing but water below. *Hornet*'s dive-bombers, bomb-laden scout planes, and fighters forged on and finding nothing, they returned home without sighting the enemy. While the Dauntless dive-bombers had enough fuel to return to either *Hornet* or Midway, the Wildcat fighters did not, and pilots had to ditch their aircraft. In contrast, Wildcats from *Enterprise* did locate the enemy carriers, but their timing did not coincide with an attack while a half dozen of *Yorktown* fighters did engage the defending Zeros. The three torpedo squadrons located the Japanese carrier force and drove home piecemeal attacks, led first by VT-8 from *Hornet*, then VT-6 from *Enterprise*, and finally VT-3 from *Yorktown*. Lumbering along in their antiquated Devastators, the American aircrews became easy prey to defending Zeros and antiaircraft gunners. Though some of the pilots successfully launched torpedoes, Japanese carrier skippers skillfully dodged the underwater missiles. Only four of the 41 torpedo planes would return. *Hornet* would not see any planes return from VT-8. Completely decimated, only one aircrew would live to witness subsequent events as Ensign George Gay kept afloat in the water among the Japanese Fleet.[15]

However, the torpedo attacks that claimed the lives of so many proved not in vain as they drew the defending Zeros to a low altitude, allowing *Enterprise* and *Yorktown* dive-bombing and scouting squadrons to arrive unmolested. *Enterprise* Dauntless bombers struck *Kaga* with four bombs and holed *Akagi* with two more. The *Yorktown*-based dive-bombers notched three hits on *Soryu*. With aircraft being fueled and armed on hangar decks below, the three ships quickly became burning infernos. By 1030, Vice Admiral Nagumo faced the real prospect that half of his carrier force that attacked Pearl Harbor six months previous would be no more. Only *Hiryu* remained.[16]

On the *Atlanta* many chafed that they could not see nor contribute to the air and sea battle raged to the west. However, that frustration quickly subsided once the dive-bombers pulled out of their dives and regrouped for the return flight home. Their reports of damage inflicted on the enemy were overheard in *Atlanta*'s radio room and the news quickly spread through the ship. Ensign Corboy wrote: "The news was so wonderfully heartening that we could hardly contain ourselves."

Not only did the *Mighty A*'s radiomen home in on the American airmen's broadcasts, but they also intercepted a Japanese air intercept control frequency. On that frequency, a Japanese air controller gave a running commentary of what was happening around the First Carrier Striking Fleet. Although no one could understand Japanese, the increasing intensity of his account, interrupted by agonized screams of "yeeeeeouch," confirmed that things were not going well for the Imperial Japanese Navy. With permission of the bridge, this continuing rambling account was piped into the ship's speaker system for the listening pleasure of all hands.

Corboy noted: "After a particularly horrible yowl from the Jap, Lt. Lloyd M. Mustin, U.S.N. drawled into the microphone 'That Jap seems to be telling Tokio [*sic*] that he's getting something and whatever he is getting, he doesn't like it a damn bit.'" In the background, *Atlanta*'s crew could hear what they believed to be antiaircraft shells exploding. Suddenly, after one final screech, the Japanese broadcaster's voice abruptly left the airwaves.[17]

Expecting a counterattack, Captain Jenkins kept his crew at General Quarters. With the news coming in over the radio circuits, the men found time to joke excessively and to complain vociferously about the quality of the lunch which the hardworking supply department had endeavored to provide: pea soup, ham sandwiches, apples, blueberry pie, and coffee. Distributed throughout the ship to the different GQ stations, the crew had to forgo using utensils. Most of the men agreed that if somebody had been around to sell peanuts and pop, the atmosphere could then well have been that of a World Series baseball game. Ensign Shaw wrote in a letter to his wife that the crew had become emotionally engaged to what was happening around them. "We were all so concerned in the success of our mission that any one word one way or the other would throw us into a frenzy of joy or a pall of gloom out of all proportion to the importance of the fact reported."[18]

The victorious aircraft that survived arrived in separate groupings and as each group arrived, Spruance's carriers turned east into the light breeze to recover. Once traps were made, the heading was again reversed to close distance on the Japanese. Because of this constant back and forth, the two task forces lost visual contact. On *Atlanta*, as the topside watch standers started eating their rations, what began as a festive meal soon turned grim as "a terrific curtain of antiaircraft fire appeared over the spot where we knew the *Yorktown* to be." Peering through their binoculars, the lookouts on the light cruiser watched as Japanese dive-bombers and torpedo planes started in and then moved away, and then attacked from another direction.

Corboy observed: "There were gigantic bursts in the sky as direct hits were made on dive-bombers. The attacking planes flew apart in thunderclaps of fire and black smoke. Streamers of smoke hung between the sky and sea as flaming planes plunged beneath the waves. There would be a flurry of bursts along the water, followed by thunderous explosions and mushrooming black pillars as direct hits were scored on torpedo planes."[19]

Atlanta's log recorded "Heavy black smoke on the horizon." Given the size of the billowing cloud, *Atlanta*'s crew solemnly concluded that *Yorktown* was "a goner." With Task Force 16 now steaming westward to close on the Japanese to launch another air attack, the cloud, marking the position of *Yorktown* fell beyond the horizon and then disappeared.[20]

However, though damaged by ordnance for the second time in a month, *Yorktown* withstood the enemy blows. Sailors quickly fought and conquered the flames and down below repairs were made and fires relit. Within two hours "*Y*" was proved capable of 19 knots and Task Force 17 attempted to catch back up to Spruance's forces.

On board *Enterprise*, Spruance considered his options. The remaining Japanese carrier remained as a threat as evidenced by *Yorktown*'s misfortune. Yet he hesitated to launch an immediate strike given he had only an approximate location based on a 1000 sighting report. Given the losses from the morning's rather uncoordinated but fortunately successful

Yorktown ablaze following an attack from aircraft from *Hiryu* on June 4, 1942. (Archives Branch, Naval History and Heritage Command, Washington, DC; NH 94451)

attack, Spruance opted to hold off until he had a better fix on the surviving Japanese carrier. Meanwhile a second wave of planes from *Hiryu* approached from the northwest.[21]

Atlanta's radar picked up the bandits some 10 miles out circling to gain an advantageous attack position. Shaw later wrote: "At 1440 a *Hornet* plane crashed near off our starboard bow and was lost at once. At 1445 the Japs were coming in fast. They disappeared in the clouds as we tracked and my spirits took a drop as director 11 reported the FD radar there inoperative." Shaw, realizing that the Japanese attackers could drop out of the clouds unseen, cursed the radar under his breath.[22]

Atlanta and Task Force 16 would be spared as the second wave of aircraft from *Hiryu* spotted *Yorktown*. Against a determined defense, two of the attackers struck fatal blows with torpedoes.

Meanwhile one of Nagumo's scout planes reported the progress of the attack, radioing back, "Friendly attack units are attacking the enemy carriers. There are three carriers."[23] This aircraft may have been one that Shaw thought *Atlanta* might get a crack at, noting one plane came within 15,500 yards, "barely within gun range." *Atlanta*'s gun barrels would remain cold, however, as Captain Jenkins determined that firing on the lone flier would be a waste of ammunition. "We were solely tempted to ..." added Shaw.[24]

Over on *Yorktown*, the onrush of water caused the ship to list to 26 degrees. With that, *Yorktown*'s Capt. Elliott Buckmaster gave the order to abandon ship. Some 3000 sailors were successfully recovered by four destroyers.[25]

With Spruance receiving an updated position report on the remnants of the Japanese carrier strike force from one of *Yorktown*'s scout planes, revenge would be swift. The two remaining American carriers turned into the wind and soon thirty-nine aircraft were in the air. At past 1700, the group assailed the remaining operational Japanese carrier scoring hits on the *Hiryu*'s bow. Within minutes, the last of Nagumo's carriers was set ablaze beyond repair.[26]

On *Atlanta*, dinner, consisting of mostly of apples and sandwiches, was once again served at the various General Quarters stations. There was excitement over the reports of the destruction of the fourth Japanese carrier, but also some disappointment that *Atlanta*'s batteries were not called into action.

The decision on how to proceed now fell on Spruance's shoulders. Earlier in the afternoon, Fletcher had passed Spruance overall tactical command as the senior admiral, having transferred his flag from *Yorktown* to the cruiser *Astoria*, realized he had limited command and control facilities in *Astoria*. Spruance retained much better situational awareness embarked in *Enterprise*.

While Spruance must have been elated by the news of *Hiryu*'s fate, he remained attuned to the Japanese maintaining overwhelming superiority in escorting battleships and cruisers that were located beyond the western horizon some 100 miles distant. Conceivably, if he continued westward at a 25-knot pace and the Japanese matched that on an eastward heading, the two forces would come within visual distance in less than two hours. While radar offered some advantages, a night engagement against Japanese battleships was simply not going to be allowed to happen.

Consequently, Spruance ordered the two task forces to the east. On *Atlanta*, the crew remained at General Quarters as evening approached. Corboy remembered the "darkness creeping up from the east and gradually blotting out the sky and clouds while in the west the sunset flamed in a riot of colors."[27] *Atlanta*'s war diary noted the ship secured from General Quarters at 2125 and that Task Force 16 returned to the 11-Victor formation with *Atlanta* assigned as the guide ship.

At midnight, Spruance turned north for an hour and then proceeded west so that they would be positioned to defend Midway in the morning against further Japanese assaults.[28] Spruance's decisions proved prescient. Admiral Yamamoto did order an eastward surge of his major combatants to draw Spruance into a nocturnal gun duel. After plying ahead in darkness for several hours through empty Pacific waters north of Midway, Yamamoto realized that no night action would be forthcoming and directed a reverse in course. Separate from this eastward thrust a surface group of four cruisers and two destroyers closed on Midway to conduct a predawn bombardment that had been planned as a preliminary to the invasion. At 0020 Yamamoto canceled the bombardment mission and two and a half hours later, he ordered the cancelation of the whole Midway operation and withdrawal northwest for a refueling rendezvous on June 7. Awake watch standers and their snoozing shipmates in *Atlanta* had yet to realize that victory had been won.

On the light cruiser, the boatswain's mate of the watch sounded reveille at 0430. Unlike the previous day, those standing watch topside were subjected to a soaking downpour as visibility dropped to 1,000 yards. With Spruance's forces now east of Midway, they were now heading westward with the hope that remnants of the Japanese Fleet could be subjected to air attack. A report of enemy forces some 50 miles west of Midway apparently identified the ships assigned to that aborted bombardment mission. With no action imminent with the threat of enemy air attack abated, Captain Jenkins stood his crew down from an extended morning GQ at about 0900. Lieutenant Mustin observed that the crew were "half asleep on their feet."[29]

With most of the Japanese Fleet having turned on a northwesterly heading, the aborted bombardment group represented low-hanging fruit for the remaining American air forces based on Midway and Spruance's two carriers. During the night the Japanese surface force inflicted damage upon itself. Reacting to the sighting of an American submarine, the column turned left to avoid an anticipated torpedo spread. *Mogami*, not responding quickly to the flashing light turn signal, rammed the stern of *Mikuma*, crushing *Mogami*'s bow and puncturing a fuel tank in *Mikuma*.[30]

As the two unscathed cruisers forged ahead at top speed to distance themselves from American aircraft, the two damaged cruisers with a destroyer escort departed the area albeit with a slower speed of advance. Midway-based aircraft sought them out. The four ships fended off an initial attack by a dozen Midway-based Dauntlesses and Vindicators. The only damage suffered resulted from Marine Cpt. Richard Fleming, who flew his damaged torpedo plane into *Mikuma*'s aft turret. Eight B-17s straddled the warships with one near miss killing two *Mogami* sailors.[31]

For the two Japanese cruisers under duress, the weather and misleading intelligence came to the rescue. Squalls continued into the early morning, making flight operations problematic. *Atlanta*'s records reported visibility reduced to one mile. During the night *Atlanta* steamed east and then south for an hour before turning west with the task force at 0400. At 0631 the course was adjusted to 230 degrees with speed increased to 20 knots. At 0719, Spruance ordered the task force speed increased to 25 knots because the task force commander heard inaccurate reports that two Japanese carriers were burning but afloat. At 1000, the two carriers commenced launching 58 dive-bombers. The sortie's search yielded nothing because *Hiryu* had passed beneath the surface earlier in the day. Coming upon the destroyer *Tanikaze*, the American dive-bombers pressed an attack, scoring no hits and lost an aircraft to the defender's guns.[32]

Rather than pursue the two crippled cruisers, Spruance lashed out one more time to catch up with Yamamoto's fleeing capital ships. Unfortunately, with the fleeing Japanese Fleet beyond the range of Spruance's dive-bombers, the American commander faced the prospect of losing more naval aviators as darkness approached with planes still in the air. Aware of the Japanese submarine threat, Spruance opted to have his carriers illuminate their flight decks to facilitate recovery being painfully aware that without his aircraft, he would have been defanged, and helpless to continue the battle should Yamamoto resume the attack. On *Atlanta*, the lookouts strained to spot periscopes or torpedo wakes. Spruance's gambit paid off. All but one dive-bomber touched down on his two flight decks.[33]

With a heavy weather front coming in from the northwest, Spruance determined that further chase of Yamamoto's fleet would no longer be likely to bear fruit so he steamed to the west with hopes of capturing *Mikuma* and *Mogami* before they too could sneak away. *Enterprise* launched an early morning scouting mission that trailed an oil slick up to its source. Subsequently, *Hornet* dive-bombers delivered blows on the two cruisers. However, since the bombs being dropped were not armor piercing, the hits merely mangled the topside superstructure with nominal impact on the vessels' overall watertight integrity.

The next wave of dive-bombers from *Enterprise* scored more hits. A third *Hornet* wave of dive-bombers against the two stricken ships also recorded more hits. On board *Atlanta* the radio room picked up the transmissions from the *Hornet's* attacking dive-bombers. Total radio circuit discipline broke down as the pilots dropped one bomb on target after another. "Look at that son-of-a-bitch burn! … Hit the son-of-a-bitch again! … Let's hit them all … your bomb really hit them in the fantail. Boy, that's swell … These Japs are easy as shooting ducks in a rain barrel."[34] Lieutenant Shaw recounted, "[T]he conversation was rich and racy with the thud of exploding bombs in the background.… [T]he whole thing sounded like a bunch of kids playing Cowboys and Indians, fantastically unreal."[35]

Mogami would survive and would see action later in the war. However, hits on *Mikuma* proved fatal. *Mogami* and the two escorting destroyers picked up many of *Mikuma*'s survivors. The heavy cruiser permanently exited the Japanese order of battle that evening with a plunge to the bottom of the Pacific.

Admiral Yamamoto hoped that the punishment suffered by his two cruisers would not be for naught. As his two cruisers fended off air attacks on the 6th, he hoped to draw the Americans within range of Wake Island-based bombers. In anticipation of Spruance continuing westward, Yamamoto maneuvered his capital ships to the south through the evening of the 6th.[36]

Task Force 16 continued to steam westward late on the 6th with *Atlanta* remaining as the formation guide, flanked by the two carriers. Just four destroyers remained screen six cruisers and two carriers against the submarine threat. Spruance had detached some of his tin cans to protect the heavily listing *Yorktown*. Recognizing that he would be coming into range of Japanese land-based airpower, conscious of his fuel state, and realizing that his sailors and aircrews had been pushed beyond physical exhaustion, Spruance reversed course.

Petty Officer McKinney wrote in his memoir: "[A]fter the engagement was broken off, we did steam northwest for some time, hoping to catch up with damaged Japanese 'stragglers,' including *Haruna, Nagato, Kirishima, Kongo, Hiei,* and *Mutsi*, all old, but their fourteen and sixteen-inch guns would have made short work of our cruisers and destroyers. We had no battleships within 3,000 miles!"[37]

Submarines also concerned Spruance who understood the Japanese had initially deployed several to set up a barrier between Pearl Harbor and Midway. On occasion during the confrontation, lookouts reported periscope sightings and the task force reacted with an emergency turn to foil any torpedo targeting solution.

As Task Force 16 came about, an abandoned *Yorktown* listed to the east. A minesweeper attempting to tow the damaged carrier on the 5th made little headway. On the morning of the 6th, the destroyer *Hammann* pulled alongside and dropped off a salvage party. Sailors hustled to put out the remaining fire, ease the list, and eliminate additional flooding. Under the direction of Captain Buckmaster, the sailors seemed to be turning the tide in the early afternoon, when *I-168* fired a spread of four torpedoes from the starboard beam. One torpedo missed. The second two passed under *Hammann* and detonated against *Yorktown*. The fourth torpedo smacked *Hammann* amidships, breaking the destroyer in half, giving few the chance to abandon ship before plunging beneath the surface.[38] *Yorktown* would follow the next morning.[39]

Having crossed the International Date Line on the 5th, entries in *Atlanta*'s war diary tracked the egress of Task Force 16 back across the date line. To save fuel, Spruance slowed to 15 knots as the formation continued eastward on June 7. The spotting of *Cimarron* early on the next day produced a collective sigh of relief as the task force formed a circular formation around the fuel ship and the thirstiest ships received head-of-the-line privileges to come alongside. Upon hearing that the oiler *Guadalupe*, with two escorting destroyers, was steaming to the west, *Atlanta* was tasked to corral the floating gas station and her attendants. While en route, *Atlanta* came upon two more destroyers—*Worden* and *Maury*. As the senior skipper, Captain Jenkins directed the two small boys to form a screen to protect against submarines.

After spotting the oiler in company with the destroyers *Ralph Talbot* and *Blue* at mid-morning, the six ships formed up and headed on a southeasterly course, rendezvousing with Task Force 16 at 1400. The rest of the afternoon was spent with the various screening ships taking turns alongside the two oilers.[40]

Fueling resumed the next day as the task force returned to the vicinity of Point Luck. *Atlanta* came alongside *Cimarron* to refuel from a near-empty vessel that looked like a floating milk carton. Sailors on deck noted that her bow was high enough out of the water to expose her keel. Having expended all of her cargo fuel, the oiler dipped into her own bunkers to provide *Atlanta* with a safe margin for the return trip to Pearl.[41]

On the morning of June 10, Task Force 16 met up with *Saratoga* having just arrived from the States with her escorts. Inclement weather delayed aircraft transfers as *Atlanta*'s log read: "Maneuvered at various courses and speeds attempting to get out of fog and rain." No flight operations occurred that day and *Spruance* turned south and kept the formation speed at 12 knots.

Anticipating the move heading north to challenge Japanese operations in the Aleutians, Captain Jenkins declared that Wednesday afternoon to be a "Rope Yarn Sunday," a Navy custom that allowed non-watch standers to slack off from their normal chores to instead work on cleaning and refurbishing their uniforms. With the portent of operations in northern climes, sailors pulled out their cold-weather gear.

The weather cleared the next day and aircraft flew off *Saratoga* to land either on *Enterprise* or *Hornet*. Spruance then turned due north to head towards the Aleutians, However, before any distance was covered, Spruance received a recall order from Nimitz to return to Pearl Harbor. In the *Atlanta*'s war diary the directive read: "Force ordered to return to Pearl Harbor." Nimitz worried that Yamamoto still hoped to engage in the "decisive battle" and he felt no obligation to provide him the chance.

Throughout the rest of the 11th through the 13th, *Atlanta* remained on a southeasterly heading towards the Hawaiian Islands. *Atlanta*'s deck log reflected slow step-ups of speed—"liberty turns"—as the cruiser got closer to Oahu. Entering the harbor, Ensign Graff's signalman hoisted the ship's two largest American flags from the fore and aft mastheads.[42]

CHAPTER 6

Intermission

Having returned to Pearl Harbor on the afternoon of Saturday June 13, *Atlanta* was reassigned to berth C-6 with her port side tied to *San Diego*, her Quincy-built sister ship. Bags of mail from the Fleet Post Office awaited and were quickly hauled aboard. The crew waited in anticipation of the magic two words—"Mail Call."

For some the postal service would not suffice as a means to reconnect with home. At his first opportunity, Lieutenant Perkins rushed to the telephone exchange to call his wife. The length of the line was such that he would not be able to return to the ship for dinner. Indeed, he would miss curfew. Finally the call was placed. After the initial exchange of mushy pleasantries, Perkins passed prepared tidbits that would not arouse the suspicions of the censor but tipped off Betts that he would be safe for a while. "You know the trouble I have with my bottom? Well I'm going to get it fixed and hope it will be better well before Nane's birthday." Betts correctly interpreted this to mean *Atlanta* would be going into dry dock to get her hull cleaned and painted and this would occur before Nane's birthday on July 12. Once Perkins completed his call, he hitched a ride with the shore patrol to a Honolulu hotel where he bedded for the night.[1]

Though Sunday was a day of rest, working parties were organized to handle the arrival of provisions. The arrival of more canned goods and other needed repair parts, paint, and lubricants, continued on Monday. However, *Atlanta*'s crew had more than enough sailors to handle these chores, allowing Captain Jenkins to place into effect a port-starboard liberty routine that allowed half of the crew to go ashore. Besides bringing aboard supplies from ashore, the remaining crew members worked on painting and cleaning the "*Mighty A*" for an inspection planned for the following Saturday. One of the signalmen noted that the ship was so clean "that one could rub against anything and not get dirty." Much to this young fellow's horror, when he went to his battle station aft of the second smokestack on inspection day and "what should I see but little black specks over the entire place." Under the cover of darkness, the adjacent *San Diego* had applied steam to her boiler tubes, knocking off accumulated soot. The hot steam rose from her stack carrying the soot into the night air, where it slowly descended onto the topside spaces of *Atlanta*. Almost as if General Quarters had been sounded, deckhands scurried topside with rags and water buckets in hand to respond to the damage of this unconventional air attack.[2]

When Captain Jenkins came topside to look over the spaces, he obviously noted what had happened. His comments to the men were that that everything was very "shipshape."

For the half of the crew that went ashore for daylight liberty, many headed into Honolulu to take advantage of the entertainment venues available. For others, there were bicycles available for trips into the countryside. One sailor found "an enchanting countryside of wildly colored vegetation dotted with the quaint homes of small oriental farmers."[3]

While some sailors pedaled around the countryside, others rode in comfort. Clifford Dunaway recalled he joined with a group of six other shipmates to rent a Model A touring sedan and venture into the tropical hills. Coming upon a pineapple farm they noted young women working in the field. Stopping the car, they hopped out to engage the ladies in conversation. However, their efforts to woo the ladies were stopped dead when an overseer came over to tell the sailors that they had wandered on to the grounds of a women's prison.[4]

Up in the wardroom there was cause for celebration as ALNAV 119 dated June 15, 1942, transmitted to the fleet the promotion list for lieutenant commanders. Lieutenants Mustin, Smith, and Wulff were notified that they were on the list. Two days later, Captain Jenkins reviewed another ALNAV that promoted his senior medical officer to commander and 10 of his ensigns to lieutenant junior grade.[5]

On the morning of June 21, after a week in port at Pearl Harbor, *Atlanta* departed with *San Diego* and spent several hours firing ammunition before returning to port in the early afternoon. Four days later, in the late afternoon, the twin ships departed again joined by the destroyers *Benham* and *Ellet*. Early the following morning, *San Diego*'s gun crews took aim at a passing drone aircraft and failed to register a hit. *Atlanta* took her turn. As the following transcript indicates, the *Mighty A* did not fare well either. The following transcript eventually published in *Chock "A" Block* recorded the sound-powered phone conversations:[6]

1000—GENERAL QUARTERS

1004—PLOT TO CONTROL: "Control, all stations manned and ready, except Turret 8 has 7 men in the brig and 1 man lashing up his hammock."

1010—CONTROL TO PLOT: "Make all tests, Plot."

1018—CONTROL TO ALL STATIONS: "Planes reported to be in the air with drone."

1019—CONTROL TO ALL STATIONS: "Radar contact, bearing 222 degrees, range 22 miles, Dir. I and II train on bearing and pick up target."

1020—RADAR TO CONTROL: "Radar has lost contact, Control." CONTROL: "Aw shucks."

1021—CONTROL TO DIR I & II "Have you picked up the target yet?" DIR I TO CONTROL: "Dir. I is on 3 planes, estimated range—" CONTROL: "All stations standby." TURRET 7 TO CONTROL: "Control, Turret 7 wishes to report a hole in the left glove of the right gun hot-case man." (Control is silent on this report.)

1022—DIR. I TO PLOT: "Standby to attack air target."

1023—DIR I TO ALL TURRETS: "Standby to commence firing, all turrets report ready." TURRETS: "Turret 8, 3, 5, 7, 2, 4, 1, 6, Aye, Aye." CONTROL TO DIR. II: "Have you picked up the target yet?" PLOT TO DIR I: "No range spot yet Dir. I."

1024—DIR II TO CONTROL: "We have a pip on the radar, but it's not strong enough to track yet." DIR II TO ALL TURRETS: "Standby to open fire."

1025—RADAR TO CONTROL: "Radar out of commission, Control." CONTROL: Gosh Darn—Control Aye, Aye."

1028—CONTROL TO PLOT: "Have you a solution yet?" PLOT TO CONTROL: "No solution, Plot." CONTROL: "What's the trouble plot?" PLOT TO CONTROL & DIR. I: "No range spot, Dir. I press your button." CONTROL AFT: "All turrets standby to fire on aircraft target." TURRET 3 TO CONTROL: "Two starshells up, Turret 3." CONTROL TO TURRET 3: "(Censored)."

1030—CONTROL TO DIR I: "Dir I designate target to Dir II."

1031—DIR II TO CONTROL: "Dir II on target, 3 planes, Plot standby to track air target. Control, I do not see the plane yet, estimated range ----."

1032—CONTROL TO RADAR: "Is the radar back in commission yet?" RADAR TO CONTROL: "Not yet, sir. We twisted one of the 40 lead cables on two, sir. Will take a few minutes." CONTROL TO RADAR: "Control Aye, Aye. Report how many leads were in that cable." RADAR TO CONTROL: "40, sir." CONTROL TO RADAR: "Sounds serious, report when its repaired." CONTROL TO PLOT: "Do you have a serious solution yet?" PLOT TO CONTROL: "No solution, Plot. No range spot yet, press your button Dir II."

1034—PLOT TO ALL TURRETS: Standby to commence firing." CONTROL TO PLOT: "Do you have a solution yet, Plot? PLOT TO CONTROL: "No solution, but we may as well have the turrets standby so when we get one there will be no delay."

1045—(Word over loud speakers): "Dinner will be served on battle stations." CONTROL TO CONTROL AFT: "I believe we'll be thru in time to eat regular dinner, what do you think Control Aft?" CONTROL AFT TO CONTROL: "Well it's a good idea, although we may get a solution before lunch time."

1046—DIR. II TO ALL STATIONS: "Standby to commence firing." TURRET 4 TO CONTROL: "Control, Turret 4 cannot see out of the periscope." CONTROL: "What's the trouble, 4." TURRET 4: "I don't know, unless the cover is on." CONTROL: Silent.

1047—PLOT TO DIR. II: "No solution Plot." DIR I TO CONTROL: "Control, those three planes turned out to be birds." DIR II TO ALL STATIONS: "Standby."

1048—CONTROL: "Report that last word, Dir I." DIR II TO CONTROL: "Dir II on same birds, Stand easy." CONTROL TO ALL STATIONS: "Here comes the drone, he's about on us, match designators, all turrets standby." DIR I TO CONTROL: "Dir I does not see the target yet." PLOT: "Range set, plot tracking."

1049—DIR II: "Director II on target, standby." CONTROL: "Commence firing when we have a solution Dir II."

1050—DIRECTOR I: "Dir I on target, Plot start tracking."

1051—CONTROL TO PLOT: "Have you a solution yet?" PLOT: "No solution yet, we had the JV and JX crossed, give us the speed and target angle again please." TURRET 5 TO CONTROL: "Permission to secure 6 ship's cooks."

1052—CONTROL TO DIR. II: "Open fire if you are on target Director II." DIRECTOR II: "All stations, standby." TURRET I TO CONTROL: Turret captain, Turret 1 requests permission to go to the head. CONTROL: "Permission granted - - WHAT!"

1053—CONTROL TO DIR II: "What are you waiting on, Dir II?" DIR II TO CONTROL: "I was on the controlling plane instead of the drone, standby to track new target, Plot, same formation." CONTROL TO DIR I: "Director I open fire when ready." DIR I TO CONTROL: Dir. I was on one of the control planes also. Plot standby to track new target, all turrets standby." TURRET 6 TO CONTROL: "No power on Turret 6, Turret 6 going off the line."

1056—CONTROL: "Cease Firing, all turrets." CONTROL TO CONTROL AFT: They ought to have bigger gas tanks on them drones so they would give us a chance to get a shot at them, the control plane reported the drone was running low on fuel, what do you think control aft?" CONTROL AFT: "It would be a good idea alright." (Word over loudspeaker): "Regular chow will be served in the mess compartments."

1100—CONTROL: "All turrets train zero and one eight zero, secure from general quarters."[7]

What unnerved Lt. (jg) Ed Corboy were the simulated attacks made by planes off *Enterprise*. "It's an exciting feeling to stand there in the director, throwing a battery around trying to get it set on a torpedo plane that is maple leafing in at dizzying speed. Your 'enemy' is 1,000 yards off. Then he's 75. Then he flips up, missing you by about 10 yards and grins down as he goes over."[8]

Healthy competition arose among the gun crews. While the machine gun crews exclaimed "They were on the ball with Ensign Hall," the 5-inch 38 crews saw themselves on the top of the pecking order. Several of those crews were led by Lt. Al Newhall. When he departed for flight school, he left Corboy with strict instructions to "keep Dave Hall envious of our boys."[9]

As Lieutenant Perkins craftily conveyed to his wife, *Atlanta* would finally get an opportunity to dry dock to replace the protective coatings of paint that had fallen away, exposing bare steel to rust and marine growth.[10]

On Tuesday June 30, the crew learned that each division would have responsibility for a certain section of the hull. The next morning, tugs eased the light cruiser and nosed her into Pearl Harbor's dry dock #2. Once inside, lines were dropped over the side to men waiting in row boats and the boats brought the rope to awaiting dockworkers who heaved to place the ship to within an inch of where she needed be above the blocks below. Once the lines were secured, long shoring planks resembling telephone poles were put in place as insurance against the ship toppling over as the water was pumped out. Looking out, the sailors could see the dock slowly rising around them; then the rising stopped as the ship came to rest on the keel blocks which was noted at 1043 in the deck log. Suddenly, the word was passed for all hands to shift over to the port side. The sudden movement of weight apparently was enough to settle the ship where she needed to be.[11]

With the cruiser firmly resting on the keel blocks, the water continued to recede, and the crew mustered at their assigned hull-cleaning stations. Scaffolds were erected on both sides of the ship extending from the bow on aft. On each section of scaffold, four-man teams, armed with scrapers and wire brushes, climbed over the side and descended down to the waterline. Each man had the benefit of being attached to a safety line. Initially, this precaution was not critical since there was water still in the dry dock to cushion a fall, however, as the void emptied later in the day, the concrete bottom loomed below.

Scraping began at noon and continued until after dark. As the *Atlanta* crewmen worked their way down, Shipyard painters took positions in the scaffolding just over their head and started spraying a coat of rust preventive. Thus, a race ensured as the crewmen worked to keep ahead of the painters. Some hull sections could only be reached using long-handled tools. By the time the water receded below the keel blocks, the painters had caught up with the scrapers and continued to paint, even over areas that had not been touched. Needless to say, between the gunk coming off the hull and the paint spraying down from above, *Atlanta*'s crew was covered in grime at the end of the day. As an added inconvenience, the crew had to wash themselves with cold water in a darkened shed located at the end of the dry dock. No showers were available in *Atlanta* due to the shutdown of the desalinization plants.[12]

Late afternoon on Thursday July 2, *Atlanta* was refloated within the confines of dry dock #2 and moved out for a temporary berth alongside *California*, which was readying for a return to the mainland for months of major reconstruction and modernization.[13]

Over the 4th of July weekend, the crew enjoyed some hard-earned liberty. Not everyone got to hit the beach right away. Prior to granting liberty on the nation's birthday, the commanding officer held Captain's Mast and assigned extra duty to those who had been late returning back to the ship over the previous few days. Those having the duty on Sunday hardly had a day of rest as eight torpedo warheads, 10 Mark VII 600-pound depth charges, 24 Mark VI 300-pound depth charges, and 200 half-pound demolition blocks needed to be hauled aboard and stowed.[14]

On the morning of Wednesday July 8, *Atlanta* departed Pearl Harbor for training exercises with *Enterprise, San Francisco, Portland, San Diego,* and others from Task Force 16. Following early afternoon gunfire exercises with *San Francisco* and *Portland* against targets towed by tugs, *Atlanta* conducted antiaircraft practice at aerial targets. Once again, practice needed to be called off following the knocking down of yet another target sleeve, this time on the second pass.[15] With *Enterprise* conducting flight operations the following day, *Atlanta*'s crew ran through several damage control problems. On July 10, *Atlanta* joined a group of destroyers assigned to shell the uninhabited island of Kahoolawe. Following the shore bombardment exercise, *Atlanta* concluded a hectic day by shielding *Enterprise* against two destroyers attempting to emulate Japanese attack tactics.[16]

While en route back to Pearl Harbor the next morning, *Portland, San Francisco,* and *Atlanta* again formed a battleline to fire broadsides at a towed target. Back at berth C-6, *Atlanta* once again nested alongside *San Diego*. The next day, a Sunday, a fuel barge discharged fuel into *Atlanta*'s bunkers, and working parties lined up to take aboard provisions for an imminent deployment. The sudden arrival of *North Carolina* likely boosted morale. With all of the attention the sleek new battlewagon had been receiving, the moniker "Showboat" seemed to stick.[17]

Though having fine lines, the new battleship really wasn't that much more powerful than the earlier classes that had been savaged seven months earlier. The big contrast was *North Carolina*'s 7 knots plus in speed advantage over her World War I-vintage sisters. Ten 5-inch twin mounts—two more than *Atlanta*—would also make this ship a formidable anti-aircraft platform.[18]

CHAPTER 7

En Route to Guadalcanal

"The action at Midway gave us breathing space to get reorganized," reflected Bob Graff decades later. Although the Japanese had received a stunning blow at Midway, their navy remained intact, potent, and capable of continuing offensive operations. As the Japanese Combined Fleet steamed towards Midway, Japanese engineers surveyed a site on the north side of Guadalcanal, an island located the southeastern end of the Solomon Islands chain in the Southwest Pacific, with the intent of constructing an airfield. From such a strategic location, Japanese aviation could seize control of the airspace over the primary shipping lanes between the United States and Australia. Also at risk—American bases at Espiritu Santo, Efate, and possibly New Caledonia.[1] As Japanese construction of the airstrip began in June and continued into July, plans evolved in discussions between Washington and Pearl Harbor on how to counter the Japanese move.

Thus, on the morning July 15, *Atlanta*, as part of Task Force 16, departed Pearl Harbor for her last time and was ordered into a circular formation. Stationed 4,000 yards off the starboard beam of *Enterprise*, the light cruiser followed in the wake of the destroyer *Benham*. The cruiser *Portland* steamed directly off *Atlanta*'s port beam, guarding the formation's left flank. Trailing 4,000 yards astern of the aircraft carrier, *North Carolina* was paralleled by two destroyers. That evening *Atlanta*'s crew had the opportunity to react to a real-time threat as built-up grease within an exhaust vent in the crew's galley caught fire. Lieutenant Commander Sears would be able to report that it took less than 13 minutes to extinguish the flames.[2]

Four days out of Hawaii, *Atlanta* crossed the equator once again. *Atlanta* had embarked a reporter, Bob Miller, from the United Press Association. Miller started his career with United Press prior to the war as a part-timer while attending the University of Nevada. He would continue working for United Press as a war correspondent, retiring in 1987. After the Crossing of the Line ceremony he wrote to his supervisor Earl J. Johnson:

> Dear Boss:
>
> Many a time I have wondered just why in Hell I ever decided to be a War Correspondent, but I never gave as much thought to the question as I did on a certain Sunday of a certain month that is famed for vacations and hot weather. That Sunday, Boss, was the day I was hailed before the court of a long-whiskered gent called Neptunis Rex, Ruler of the Raging Main, and charged with being a Pollywog.
>
> Do you know what a Pollywog is, Boss? Well, if you ever cross the Equator on the Mighty "A" you'll certainly be shocked to find out. I was a Pollywog, Boss, until that fateful Sunday, and then after a little ceremony on the fan-tail, of which I was, unfortunately, one of the central characters, I became a Shellback. Boss, you'll never know what a great thing it is to be a Shellback.
>
> Now I can't tell you just what happened during the 24-hours that preceded my debut as a Shellback as all the doings were mighty secret. But just to give you a rough—and I mean rough—idea of what happened, let the fact sink into your mind that there were 15, three more than a dozen, of us miserable pollywogs and hundreds of Shellbacks.
>
> Of those Shellbacks, four-fifths of them were neophytes, that is they had only been Shellbacks for a few weeks and still had some nice bruises, and some rather resentful memories of their recent transformation from Pollywogs to Shellback and, Boss, they were just dying to get in a few whacks to compensate them for their own injured feelings.
>
> There's no truth, Boss, to the report that I tried to get the Captain to turn the ship around and head back north when I found what was coming. The truth is that I've always been allergic to tropical diseases and figured it would be much more healthful for me north of the Equator. The Captain mentioned something about somebody being a pest and headed the ship straight to the Equator and my doom. I found out later he was a Shellback.
>
> In fact, Boss, I didn't get any sympathy from anybody, I never saw such a heartless bunch in my life. I spent two whole days cultivating the friendship of the ship's Doctor and bought him nearly two books worth of gedunks, hoping I

> could prevail on him to shoot a little pain killer into certain parts of my anatomy that were going to be more or less exposed to the elements, but he said they were all out of morphine. Yeah, Boss, he was a Shellback, too.
>
> We'll dispense with the preliminaries. They were bad enough but nothing in comparison to what happened between the hours of 9:32 and 11:16 o'clock that Sunday morning. It was exactly 9:32 when the resounding whack fell upon the skin-tight and water soaked whites covering my backsides, and it was about 11:16 that they carted me off in a stretcher considerably much worse for wear.
>
> As I write this propped up in bed in the sick bay, swathed in bandages and insulated by several pillows, I want to pay my respects to the surgeon. They say he did a magnificent job in pulling me through, and although you may have difficulty in recognizing me as a result of his surgery, you should be thankful you still have a correspondent able to navigate, however slow and painful.
>
> Yours truly and very painfully,
> Bob Miller[3]

As Miller recovered from the morning festivities, Jim Shaw noted in his diary that Captain Jenkins had announced the ship's destination as "Tonga Tabu" and that all hands posed the question "why?" Not informed of the forthcoming operation, Shaw pulled out a chart and noted that the island straddled the American "line of communication" with Australia, and looking to the west he noted the Solomons and the island of "Guatacanal [*sic*]."

During this time frame, Shaw stood watch in gun control in a Condition 3 watch section with Ens. Dave Hall and three enlisted sailors. "Over the main battery telephone we have debates, long discussions, wisecracks, and the like." Shaw argued that the banter helped relax the watch standers and made them more alert. In contrast, inside the pilothouse, Captain Jenkins kept a tight rein on his officers of the deck and bridge watch teams to assure *Atlanta* stayed on her assigned station, and zigged and zagged in the proper sequence. One of the OODs now, Lieutenant Commander Mustin recalled the atmosphere on the bridge stayed serious.

Writing to his wife, Van Ostrand Perkins noted that many of the officers had bonded and developed friendships. In addition to his friendship with Shaw, which preceded his tour on *Atlanta*, he enjoyed sharing meals with Lt. Carl Garver and Lt. (jg) Robert F. Erdman, two of the ship's medical officers. He did express some annoyance over some of the reserve junior officers who lacked the training and refinement offered by the Naval Academy. His disdain for those without an Annapolis background was not universal, as he mentioned one of his closest friends was Bob Graff, a reserve officer with a degree from Harvard.[4]

On the 24th, or rather the 25th having crossed the International Date Line, *Atlanta* arrived at Nuku'alofa, a small town that served as the port for Tongatabu, one of the Tongan or Friendly Islands. Located just west of the International Date Line, Tongatabu would serve as an idyllic sanctuary for thousands of American servicemen during the war. However, during this visit the men would be only able to enjoy eyeball liberty. Bob Graff recalled, "It was such a beautiful place … a little white Christian church and palm trees waving." Trees lined much of the remaining beach surrounding the harbor.[5]

Once anchored, *Atlanta*'s boatswain's mates turned to lowering the starboard motor whaleboat in the water which proceeded to scurry about on assorted chores. *Maury* came along the starboard side to refuel, and as a bonus, receive freshly baked loaves of bread from *Atlanta*'s bakery. Later that day, having given away some of her fuel reserves, *Atlanta* made up for the deficiency and then some by pulling alongside the tanker *Mobilube*.[6]

In addition to fuel, *Atlanta* took on some fresh stores including bananas. Petty Officer McKinney upped his reputation with his shipmates by using the fresh tropical fruit and the powdered ice cream mix on board to concoct "a passable patch of banana ice cream." The product sold well at the ship's "gedunk" stand located down on the second deck across from the Repair II damage control station.[7]

As the sun was setting, the ships departed. A delay in hoisting the #1 motor whaleboat slowed *Atlanta*'s departure. Captain Jenkins could be overheard cussing in the pilothouse. *Atlanta* was the last ship to clear the harbor.

Past noon the following day, masts were spotted on the horizon in anticipation of a giant rendezvous between Task Forces 16, 18, and 44. The latter combined American and Australian cruisers and destroyers escorting transports carrying regiments of the 1st Marine Division. The merger brought some 72 ships together at a location some 400 miles south of the Fiji Islands. The objective: the seizure of Guadalcanal and the adjacent island of Tulagi in what was dubbed Operation Watchtower. Vice Adm. Frank Jack Fletcher, the recently promoted veteran of Coral Sea and Midway, assumed overall command of this force and flew his flag in *Saratoga*. Rear Adm. Leigh Noyes commanded the Air Support Force which in addition to *Saratoga*, included *Wasp* and *Enterprise*. Escorts included the battleship *North Carolina*, five heavy cruisers, *Atlanta*, 16 destroyers, and three oilers.[8]

The formation of ships proceeded to the Fiji islands where amphibious forces conducted a three-day rehearsal landing along the northern coast of Koro Island. *Atlanta* would not be there to support the mock invasion, instead steaming well offshore to protect the carriers. The rehearsal did not go well, and the American commanders could only hope that the actual landings would go much better. On August 1, the forces regrouped and headed on a deceptive feint towards the Coral Sea. Until August 4, few onboard *Atlanta* knew about the objective. That changed.

On August 5, Shaw wrote in his diary:

> Our objective IS the Solomons, specifically the port of Tulagi, the islands of Florida and Guadalcanal. The Japs have built themselves an airbase there and established some two thousand troops. We have about fifteen thousand troops. The Japs have about 150 planes in the area are readily available against about three hundred for us. They have four or five heavy cruisers and a squadron of destroyers. It would appear, off hand, that we possess numerical superiority in every sense. I'm a little skeptical about this being a push-over tho![9]

The next day, on the eve of what would prove to be a long struggle, *Atlanta* published the first edition of *Chock "A" Block*, the ship's monthly newspaper. Run off from a mimeograph machine, the 17" × 9 ½" multi-page journal featured a header with a sketch of the cruiser surging forward creating a mighty bow wave. The crew of Turret 2, Condition 3-Mike, Watch 2 came up with the name of the paper.

The lead article was a message from Captain Jenkins. It read:

> A lot of water has gone over the dam since that cold raw day in December, 1941, when we all stood in the rain to witness the brief ceremony in connection with the commissioning of the ATLANTA. At that time, being a newly commissioned ship, we were in a state of disorganization and it has only been through the hard work of all hands that order and efficiency has come out of what might be

called chaos. After seven months of drills and exercises I now have the utmost confidence in the ability and desire of the crew to fight this ship efficiently to the limit of its endurance.

Some of you, as in my case, lost former shipmates in the Pearl Harbor attack. This fact, aside from the political issues involved in this war, spurs us on and makes the monotony of almost endless days of cruising at sea worthwhile knowing that by our efforts we are contributing toward the destruction of Japanese sea power and avenging our friends who were stabbed in the back.

Until recently, the Japs have had the initiative, now we are taking the offensive and it may mean losses to our Fleet but the advantages gained will be in our favor.

Let's keep the ship headed westward and the chin up![10]

CHAPTER 8

Cactus and Combat

D-Day, August 7, started early in *Atlanta*. The chow line opened at 0400. With the crew fed, the ship went to General Quarters. It was still dark, but topside sailors looked over at the three carriers launching their F4F Wildcats and SBD Dauntlesses aircraft into the wind. Ensign Corboy described it as a magnificent sight "as our planes began taking off in the luminous darkness of half-dawn. All our ships switched on their red mast lights and the planes turned on their flying lights as they left the carriers."[1] At 0622, *Atlanta*'s lookouts spotted the objective to the east as a glimmer of light caused the dark silhouette of Guadalcanal to come visible on the horizon. With first light, air raids against Guadalcanal and Tulagi caught the Japanese by surprise, sinking the Japanese seaplane squadron at Gavutu, a tiny speck of land in protected waters on the southern side of Tulagi.[2]

As the Marines prepared to land in the early morning, the gunfire of the cruisers *Quincy, Vincennes, Astoria,* and *Chicago*, and destroyers stationed closer, tore into the jungle, ripping up trees and foliage.[3] The Marines then stormed ashore unopposed. Indeed, during the first day ashore the struggle focused on overcoming the dense vegetation to reach the objective—the near-complete airstrip. Likewise the landing on Tulagi met little initial resistance, though the small Japanese garrison attempted a counterattack that evening.

Throughout the 7th back in *Atlanta*, the crew remained at their battle stations, watching the carriers recover, reload, and again launch their aircraft. To keep the crew apprised of the situation, Captain Jenkins allowed the aircraft radio frequency to be broadcast on the ship's speakers. Jim Shaw recalled, "… we could hear the fighters and bombers talking over their radios and sometimes hear the whine of engines in power dives but that was as close as the battle came to us."[4]

Past midday, the radio conversations grew intense as the first of two waves of aircraft from Rabaul made their way over a waterway passage down the Solomon Islands that would be dubbed "the Slot." The first wave of aircraft 24 Betty bombers escorted by 17 Zero fighters arrived at 1315 and failed to score any hits on the invasion force, thanks to a pesky defense put up from Wildcats as well as some Dauntlesses off the three American carriers. Nearly two hours later, nine unescorted Val dive-bombers arrived and were met by fighters off *Enterprise*. Only one of those aircraft would make it back to Rabaul, however, the attack managed to land a bomb hit on the destroyer *Mugford*, killing nineteen sailors.[5]

As the Marines settled on Guadalcanal for their first night ashore, *Atlanta* steamed through the night with Condition 3 set throughout the ship. At 0530, the ship went to General Quarters as the arrival of daylight permitted the three carriers to resume flight operations.[6] Ashore, the Marines resumed their forward movement and by late afternoon seized the airstrip as well as the construction camp and depot with trucks, a miniature railroad, construction equipment, an electric generator, and food, including bags of rice, that would all aid the American cause. The Japanese construction workers and nominal naval infantry force melted into the jungle west of the Lunga River. Across the Sound, Marines had captured Tulagi.[7]

Had Japanese scout planes located Fletcher's carriers that day, *Atlanta*'s gunners finally would have fired broadsides in anger. Instead, 23 Betty bombers, nine Vals, and a fighter escort, went after the invasion fleet. Once again, *Atlanta* followed the chatter of emanating from the American cockpits as Wildcats tore into the enemy formations. *Atlanta*'s counterparts in the Quincy-built sister ship *San Juan* were able to bring their 5-inch batteries to bear against the low-flying Bettys, taking down five of the attackers. However, the Americans afloat did not escape unscathed as a torpedo hit on *Jarvis* killed 15 sailors and the transport *George F. Elliott* sustained what proved to be fatal damage after a Japanese pilot steered his damaged bomber into that ship's superstructure.[8]

Reluctant to expose his carriers to a Japanese land-based bomber attack, Fletcher had moved eastward, steaming off the northwestern cape of San Cristobal, the next island the Solomons chain. Operating his carriers independently, the recently promoted vice admiral divided his battleship, cruisers, and destroyers between the three floating airfields. Ever cautious, Fletcher retained two-thirds of his fighters for combat air patrol, allowing approximately two dozen to fly cover for the invasion force off Guadalcanal.

A notable entry in *Atlanta*'s war diary occurred at 1804 on August 8: "Completed flight operations, set Condition 3 and formed Cruising Disposition 1 on course 140 degrees (T) at 15 knots."[9] Having recovered the last of his aircraft, Fletcher followed through on his previously stated intent to withdraw from the combat zone and turned his force, which included *Atlanta*, to the southeast.[10]

Thus, Jenkins and his crew would be well clear that evening when a Japanese cruiser force commanded by Vice Admiral Gunichi Mikawa slipped down the sound and delivered a stunning blow to a defending Australian and American naval force. Aided by the miscommunication and misinterpretation of several sighting reports and covered by rain squalls, a Japanese column of seven cruisers and a destroyer, first launched torpedoes against the cruiser *Chicago* and Australian cruiser *Canberra*, and when Japanese aircraft dropped flares to silhouette the two warships, they opened up with their gun batteries. The impact of Long Lance torpedoes and armor-piercing shells on the two Allied warships guarding the passage between Savo Island and Guadalcanal was devastating. The Australian cruiser may have fired a few rounds in response but was reduced to a flaming wreck in less than four minutes. Meanwhile, a Japanese torpedo tore into the starboard bow of the American cruiser which also received blows from Japanese shellfire. However, *Chicago*'s secondary batteries were able to land a shell on the light cruiser *Tenryu* before Mikawa's force turned to engage the American cruisers stationed to the northeast of Savo Island.[11]

At 0150, searchlights from Mikawa's cruisers illuminated *Vincennes, Quincy*, and the heavy cruiser *Astoria*. Shells followed the light beams and ripped into *Astoria* and *Quincy* as the latter warship took on three torpedo hits that led to her quick demise. Gunfire also raked the superstructure of *Vincennes*, and two Long Lances found their mark. As Mikawa's flotilla sped away to the northwest, *Quincy* succumbed to the sea at about 0235. *Vincennes* went under just before 0300. *Canberra* was scuttled at 0800 by a torpedo from the destroyer *Ellet*. *Astoria*'s crew fought to keep her afloat but a magazine explosion at 1100 sealed her fate. Early in the afternoon the cruiser rolled over to port and began a descent to what became known as Ironbottom Sound.[12]

In summary, the Americans and Australians lost four cruisers, with damage inflicted on a fifth cruiser and two destroyers. The full scope of the setback took a few days to reach down to the *Atlanta*'s wardroom and mess decks. Lieutenant Commander Mustin recorded that Captain Jenkins showed him the message on the 12th, three days after the battle.[13]

Reflecting on the outcome of the battle, Petty Officer McKinney observed "… the cold hard facts are that an aggressive, well-manned, and well-armed Japanese surface force steamed in and out of our disposition between Guadalcanal, Tulagi, and Savo and administered the most fearful, shameful, and complete beating our Navy has ever received in wartime." McKinney noted that while the U.S. Navy lost more ships and men at Pearl Harbor, the circumstances were different since that was a sneak attack in peacetime. Summarizing his disgust, he later wrote:

> It was easy for a then third class petty officer, who was a hundred miles away on a ship not engaged to pontificate and point a grim finger at questionable leadership, planning, and communications. It is more charitable to observe that we paid a horrible price to learn our enemy was well equipped, resourceful, and highly skilled in night naval maneuvers. That the forces Admiral Mikawa did not keep up their 'good' work that night and sink anything and everything American and Australian in that part of the Pacific can be attributed to a curious Japanese tendency to hesitate at the brink of total victory and settle for less.[14]

As McKinney noted, it was not as complete a victory for Mikawa as it could have been. He failed in his objective to attack the invasion force transports.

With American destroyers sweeping the waters off Savo Island recovering survivors, Fletcher's carriers continued on southeasterly heading until the mid-afternoon, then turning east and increasing speed to 18 knots. During the evening Fletcher's task force turned south then southeast at 15 knots to rendezvous the next day with the oilers *Platte* and *Kaskaskia*. The meet up came later that day. *Atlanta* waited as other ships of Task Force 16 took turns alongside the oilers steaming on a westerly heading. Finally, during the evening meal hours, *Atlanta* spent ninety minutes alongside the oiler *Platte* which pumped some 250,000 gallons of black oil into the light cruiser's bunkers. *Atlanta*'s war diary recorded: "Completed fueling at 1925 and took station 3000 yards ahead of *Enterprise*." One of the pleasant surprises that came with the fuel was mail. Morale took a boast as bags of letters and magazines were hoisted aboard and distributed. In addition, after being screened by junior officer censors, bags of return mail left the ship to eventually arrive stateside to be delivered to the families of the more than 600 crew. Over the next two days, *Atlanta*'s deck log noted various course changes aimed to keep the carriers well clear of enemy aviation as ship escorts took repeated turns topping off their bunkers through underway replenishment.[15]

Following General Quarters mid-morning on August 9, *Atlanta* began a seemingly endless patrol as an element of one of three forces assigned to protect some 600 miles of sea lanes between a base recently established at Espiritu Santo and Guadalcanal. Signalman Striker John W. Harvey noted during this period in mid-August that: "About once a week the three forces rendezvoused with a lone tanker. The forces combined while fueling operations proceeded. The tanker fueled ships on both sides of her, fueling them one after the other. In a day the three forces would be completely refueled, leaving the tanker sitting high in the water."[16] *Atlanta*'s war diary recorded that the refueling occurred more frequently with hookups with *Cimarron* on the 13th and 15th, and *Kaskaskia* on the 18th.[17]

Using leftover Japanese equipment to finish the job, Marine engineers continued to finish the runway while Marine infantrymen dug in a defense perimeter. The first flight of Marine aircraft arrived on August 20.[18] With Henderson Field (named for Major Lofton Henderson who

In this snapshot taken from *Enterprise* late in the afternoon of August 21, landing operations continue on *Saratoga* in the distance. *Atlanta* is on the right, with two heavy cruisers and some destroyers in the distance. (Archives Branch, Naval History and Heritage Command, Washington, DC; 80-G-74487)

was lost defending Midway) becoming operational, the Americans could rule the daytime skies in the vicinity of Guadalcanal and strike out at Japanese bases in the region. The airfield restricted Japanese efforts to reinforce their garrison to nighttime runs by fast destroyers. Moving with a sense of urgency, the Japanese deployed the Combined Fleet to the region. Now the Marines and supporting naval forces not only faced shore-based air threats, but also carrier-borne naval aviation from *Shokaku* and *Zuikaku*—the two remaining carriers from the attack from Pearl Harbor.

Knowing of Admiral Yamamoto's imminent arrival with the Combined Fleet and its carriers, Rear Adm. Raizo Tanaka initiated another resupply mission on the 23rd. A Catalina patrol plane spotted Tanaka's convoy coming down the Slot at 0950, and that news was noted in *Atlanta.* However, Fletcher chose to head south rather than close on Tanaka's ships. In defense of Fletcher, he acted on reports from *Enterprise* aircraft of three Japanese submarines heading southwards towards him.[19]

The entry made in *Atlanta*'s war diary at 1500 read: "Enemy convoy reported to northwest. SARATOGA and ENTERPRISE launched search planes and striking group but failed to make contact." Fletcher, finally acted on additional reports from the shadowing Catalina and at 1510, *Saratoga* launched 31 Dauntlesses and six Avenger torpedo planes towards Tanaka's flotilla. Tanaka, aware of his shadower, wisely reversed course to evade both the *Saratoga* air strike as well as nine dive-bombers and 12 fighters launched from Guadalcanal.[20]

Another scout plane spotted the cruisers and destroyers of Vice Adm. Nobutake Kondo's Advance Force steaming ahead of the main body of the Combined Fleet, which included *Shokaku* and *Zuikaku.* To keep his carriers from being spotted, Vice Adm. Chuichi Nagumo turned his main body on a northerly course at 1825. With no carrier sightings and a report placing Japanese carriers at Truk, Fletcher had no reason to believe a carrier duel was imminent. Thus, Fletcher directed *Wasp* and her escorts to rendezvous with the oilers.

Thus, on the morning of August 24, Task Force 61—consisting of Task Force 16 centered around *Enterprise* and Task Force 11 led by *Saratoga*—steamed westward off the northeast coast of San Cristobal, an island 40 miles southeast of Guadalcanal. Fletcher sought to close the distance to Henderson Field to recover his aircraft that had landed there to refuel following the failed attempt to intercept Tanaka's convoy.

Finally, at 0935, a Catalina spotted the Japanese light carrier *Ryujo* escorted by a cruiser and two destroyers. The four-ship formation was steaming ahead to provide air support for Tanaka's supply convoy, but more importantly, to bait Fletcher to attack and thus, draw a strong counterpunch from the two Japanese heavy carriers. Fletcher could not take the bait, in part because he still had inbound aircraft returning to *Saratoga* from Guadalcanal.[21]

With the *Saratoga*'s aircraft having returned during the late morning, at 1212 *Atlanta* turned with *Enterprise* to a due north heading. As the helmsman steadied up on the new course, Wildcats from *Saratoga* spotted a lumbering four-engine Japanese flying boat over the horizon and dispatched the unwanted intruder. Fletcher then ordered *Enterprise* scout planes to fan out across a northerly arc. *Atlanta*'s log noted: "Increased speed to 18 knots and maneuvered with ENTERPRISE launching planes."[22]

The following entry read: "Japanese plane sighted to northwest, went to General Quarters. Plane shot down by own fighters." Arriving at their GQ stations, topside personnel witnessed smoke emanating from what proved to be a land-based Betty bomber. While everyone assumed at least one of the two Japanese aircraft reported their presence, both had been shot down before either could get off a contact report. Unaware of the presence of the American carriers, *Ryujo* launched fifteen Zeros and six Kates to attack Henderson Field.[23]

The airborne strike could be seen some 100 miles distant thanks to *Saratoga*'s air search radar. Plotters tracked the formation and quickly discerned it was heading towards Guadalcanal, and determined the direction of the launch point. At about 1400, *Saratoga* launched thirty Dauntlesses and eight Avengers against the Japanese light carrier and her escorts. Fletcher fell for the Japanese ploy.[24]

At 1431, *Atlanta*'s war diary recorded: "Observed another plane shot down to westward" which proved to be a floatplane launched by the cruiser *Chikuma*. Suspecting that his scout had been shot down by American carrier aircraft, Nagumo now had a point over the horizon to aim strike groups from *Zuikaku* and *Shokaku*. At approximately 1500 and an hour later, two attack groups totaling 73 aircraft flew off the two big-deck carriers. While scout planes off *Enterprise* confirmed *Ryujo*'s location, another pair found the two big-deck carriers and dove on *Shokaku*, scoring near misses. Attempts to redirect the *Ryujo* strike failed. To his credit, deft shiphandling enabled *Ryujo* to evade the first 10 bombs dropped but others scored hits as did one, possible three torpedo hits from *Saratoga's* Avengers.[25]

As *Ryujo* evaded attack, Task Force 61 readied for the Japanese counterstrike. *Atlanta*'s 1530 war diary read: "Set course 335 degrees (T) at 20 knots to close reported enemy carrier group." At 1602, *Enterprise*'s radarmen detected the enemy air formation bearing 320 degrees at 88 miles. Thirty-five minutes later *Atlanta*'s war diary indicated: "Unidentified planes reported bearing 300 degrees (T). Changed course to 100 degrees (T) and increased speed to 25 knots." By this time Japanese aircraft were within 25 miles of *Enterprise*. During the time between the *Enterprise* radar sighting and the 1637 *Atlanta* war diary entry, flight deck crews on both *Saratoga* and *Enterprise* hustled to launch 53 Wildcats to intercept the incoming Japanese. Additional aircraft were launched to go after Nagumo's carriers.[26]

The Wildcats took a heavy toll on Japanese aircraft that had spotted *Enterprise*. Due to earlier separate maneuvers by the two American carriers to conduct flight operations, about 10 miles separated *Enterprise* and *Saratoga*, with the former positioned northwest of Fletcher's flagship. To defend *Enterprise*, Screen Commander, Rear Adm. Mahlon Tisdale, positioned *North Carolina* aft of the carrier and formed a circular formation with *Portland* off the carrier's port bow and *Atlanta* off the starboard bow, and the six destroyers taking other stations around the screen. The formation placed Jenkins's ship on the opposite side from where the enemy's attack was emanating, but the XO, Commander Emory, from his position aft at Battle II, enjoyed a panoramic view of the combat.[27]

Surviving a gauntlet of Wildcats, Japanese Val dive-bombers intentionally overshot *Enterprise* as well as *Atlanta* and turned to descend at seven-second intervals. With high haze and late afternoon sunshine, lookouts on *Atlanta* finally spotted the diving planes over the port bow, and the light cruiser opened up with barrage fire aimed to burst shrapnel over the sky above the carrier. With *Enterprise* first turning hard left to execute an S-turn, *Atlanta*'s helmsman followed with the hard left and then reversed to a hard right. As the formation turned, the relative position of *Enterprise* shifted from *Atlanta*'s port quarter to port beam, unmasking the three forward gun 5-inch dual mounts controlled by Lieutenant Commander Nickelson, and enabling the aft three mounts operated by Lieutenant Commander Mustin to engage, as well as the port side 20mm and 1.1-inch guns. "It was a full deflection shot as far as we were concerned." So close was the range that Lieutenant (jg) Craighill in charge of the forward machine guns simply designated targets and his pointer and trainer put their sights on the attackers.[28] With concentrated fire from *Enterprise*, *Atlanta*, and other ships in the formation, the determined Japanese pilots pressed on to what was becoming a suicide mission. *Enterprise*'s evasive action thwarted the first bomb-dropper, and the second and third planes crashed near the flattop. Other Vals came down in pieces with burning fuel following. Unfortunately, when *Enterprise* reversed rudder, the momentary rudder shift effectively straightened the ship out, making her an easy target.[29]

Atlanta's war diary recorded: "Observed bomb hit on flight deck aft. There are several near misses and at least five enemy planes shot down." The first bomb exploded in the petty officers' mess several decks below, killing 35 after piercing the flight deck on the forward right corner of the number three elevator. A second bomb landed 15 feet outboard of the first, killing another 33 sailors who were positioned at a 5-inch gun tub. Fortunately, the third bomb that landed two minutes later forward near the number two elevator caused no fatalities. Corboy would write, "It was infuriating for us to watch that big, beautiful ship go lunging about the sea with flames and smoke pouring out of her."[30]

A half hour after the bomb hits *Atlanta*'s war diary recorded: "ENTERPRISE reported fires under control." Despite the damage, *Enterprise* managed to recover nearly all of the fighters and other aircraft that were running out of fuel. *Atlanta*'s war diary noted the exceptions: "Two F4Fs made forced landings close aboard during evolution, one plane crashed on deck, another into side of carrier."

The American counterattack that lifted off from the two carriers just prior to the Japanese arrival failed to find the big-deck carriers. Two Dauntlesses off *Saratoga* found the seaplane carrier *Chitose* and with near misses, cracked hull plates that flooded machinery spaces and forced the ship to head to Truk for repairs.[31]

As darkness fell, the two opposing carrier forces disengaged and recovered returning aircraft. *Atlanta*'s war diary recorded: "Six ENTERPRISE planes approached. Maneuvered to recover but second plane crashed on deck and it was necessary to send remaining four to SARATOGA."[32]

Throughout the evening, the Task Force 11 and 16 combo that formed Task Force 61 maintained a southerly heading, and early morning the next day *Atlanta*'s lookouts sighted the oilers *Platte* and *Cimarron*, with their escorts *Clark* and *Gwinn*, on the horizon. During the rest of the day the Task Force 16 flotilla centered on the wounded *Enterprise* and replenished their bunkers. During *Atlanta*'s 70-minute hook up alongside *Cimarron*, the task force reversed course from a southern to a northern heading requiring a 150 degree course adjustment. To complete the turn, the two ships made incremental course adjustments with *Cimarron* maintaining speed, and Jenkins and his bridge team adjusting speeds to maintain position abeam.[33]

That evening *North Carolina*, along with *Atlanta*, and the destroyers *Monssen* and *Grayson*, were detached from Task Force 16 to shift to Task Force 11 as *Enterprise* departed with some escorts for a repairs period at Pearl Harbor. Just after midnight on the 26th, *Atlanta* met up with Fletcher's task force and was steaming back north.[34]

Though the visual evidence from the perspective of *Atlanta*'s crew indicated that the U.S. Navy had suffered a setback in the eastern Solomons, the Japanese strategic objective to remove the U.S. Marines from Guadalcanal and assume control of Henderson Field was not met. Rear Admiral Tanaka again steamed southeast with cruiser *Jintsu* and five destroyers to escort the transports *Kinryu Maru*, *Boston Maru*, and *Daifuku Maru* carrying 1,500 troops and vital supplies.

In the predawn moonlit darkness of August 25, a Catalina spotted Tanaka's flotilla. With first light, eight dive-bombers and an equal number of fighters were airborne to intercept the convoy located 180 miles up the Slot and closing. *Jintsu* suffered a hit as did the *Kinryu Maru*.[35] Smoke from that blazing cargo ship served as a beacon for a flight of B-17s which scored a rare high-altitude hit on the destroyer *Mutsuki*.[36] With the loss of a destroyer, cargo ship and damaged cruiser, Tanaka chose to turn away from Guadalcanal. In retrospect, due to these subsequent events, the Battle of the Eastern Solomons proved to be a greater American success than recognized at the time.

The Battle of the Eastern Solomons was significant for a second reason. For the first time since she had commissioned almost exactly eight months prior, *Atlanta* fired her guns against the enemy. Mustin recorded that *Atlanta* expended "264 rounds 5-inch, 2,000 + of 1.1-inch, and about 1,100 20mm." Along with the *North Carolina*, *Atlanta* possessed the most advanced antiaircraft weaponry in the fleet. However, the modern battleship offered a more attractive alternative target for two groups of

USS *Enterprise*'s (CV-6) bulged flight deck structure, resulting from a bomb that exploded below during the Battle of the Eastern Solomons, August 24, 1942. Photographed a few days later, after the ship had returned to port. Note the *Atlanta* (CL-51) is in the background. (Archives Branch, Naval History and Heritage Command, Washington, DC; 80-G-K-412)

seven dive-bombers. One group of three Vals dove from off the starboard bow, while the other four came in from the port quarter. A hail of rising steel swatted down each of the attackers.[37]

In contrast, Nickelson and Mustin from their respective forward and aft perches in *Atlanta* tried to shoot down aircraft diving on another ship—*Enterprise*. In doing so they gained an appreciation and the limitations of the technology they were entrusted with.

Atlanta's after-action report, sent out on August 27, tallied 264 5-inch, 2,036 1.1-inch, and 1,185 20mm rounds expended. Five sailors suffered minor injuries, mostly due to burns from expended shell casings.

Commander Emory concluded that this was the "First opportunity to join the enemy in battle" which was met by all hands "with enthusiasm."[38]

CHAPTER 9

Attrition

At 0645 on August 29, *Atlanta*'s war diary noted "Sighted Task Force 17 to eastward." Commanded by recently promoted Rear Adm. George Murray, who happened to be Lieutenant Commander Mustin's stepfather, Task Force 17 consisted of the carrier *Hornet*, the cruisers *Northampton, Pensacola, San Diego,* and five destroyers. Since playing a key role at Midway, the carrier had been sidelined at Pearl Harbor to receive several upgrades.[1]

On the following day, *Atlanta*'s war diary reflected the shuffling of the ships:

> Task Force 61 Op. Order 4-42 became effective and ATLANTA became a unit of Task Group 61.1 consisting of Vice Admiral Fletcher in SARATOGA, NORTH CAROLINA, MINNEAPOLIS, NEW ORLEANS, PHELPS, DEWEY, FARRAGUT, MACDONOUGH, WORDEN, MONSSEN, and GRAYSON. HMAS AUSTRALIA, and HOBART plus PATTERSON and BAGLEY left disposition for new assignment.

The two Australian warships were reassigned to Task Force 18 to reinforce the screen guarding the carrier *Wasp*, and the two American destroyers assumed similar missions to reinforce Rear Admiral Murray's screen around *Hornet*.

Subsequent war diary entries over the next nine hours recorded *Atlanta* changing station off *Saratoga* four times as the carrier maneuvered to conduct flight operations.[2]

On morning of August 31, the Task Group was located approximately 260 miles southeast of Guadalcanal when at 0655 the formation changed course and axis to 140 degrees (T) and slowed to 13 knots to enable the escorting destroyers the optimal speed to attain sonar underwater contacts. This new heading placed *Atlanta* nearly directly east of *Saratoga* at a distance of 2,500 yards. On his way up to the bridge, Lieutenant Commander Mustin heard a torpedo alarm on the TBS voice radio. "I just about had enough time to look up and see an enormous pillar of spray rise into the air alongside *Saratoga*." *Atlanta*'s war diary duly noted at 0747: "Observed torpedo hit on *Saratoga*'s starboard quarter."[3]

The torpedo pierced the hull of a ship that had been initially designed as a battlecruiser and flooded a boiler room that had been shut down and unmanned. However, the jolt of the blast jarred the carrier's turboelectric machinery which drove the shafts, forcing *Saratoga* to come to a slow stop. The source of the spread of torpedoes aimed at the American carrier would later be identified as *I-26*. With the Japanese submarine attacking from the far side of the formation, the next ship in line of the two torpedoes was *Atlanta*. Lookouts spotted one torpedo pass ahead and saw another pass by astern. The light cruiser's speed increased to 20 knots as Captain Jenkins steered his ship on various courses to guard the damaged carrier.[4]

Though damaged below the waterline, *Saratoga* with a stiff breeze over the bow, could still launch and land aircraft while under tow and at slow speed once some power was restored. Thus, while *Saratoga* had to depart for repairs, the air group could fly off for distribution to *Wasp*, Espiritu Santo, and Guadalcanal to join with aircrews from the recently departed *Enterprise*.[5]

Late on September 1, *Atlanta*'s war diary noted the arrival of the salvage tug *Navajo* and destroyer *Laffey*. A day later, the oiler *Guadalupe* arrived to enable Jenkins's ship to refuel the following morning.[6]

The September 6, 0735, *Atlanta*'s war diary entry read: "Sighted Tongatabu bearing 115 degrees (T) distant 15 miles." Just before noon, *Atlanta* anchored at berth 31 off Nukuʻalofa, and supply boats began coming alongside with needed provisions. Also a launch from *Atlanta*'s Kearny sister ship came alongside carrying a wardroom contingent seeking to confer with their more seasoned counterparts on operating in the Southwest Pacific. Along with *Juneau*, *Atlanta*'s sailors gawked at the battleship *South Dakota* which added nine more 16-inch guns to the Navy's front-line arsenal.[7]

The scuttlebutt had *Atlanta* heading back to Honolulu, assigned to escort *Saratoga*. However, that afternoon, *South Dakota* steamed over a coral reef that supposedly had been cleared, gashing a hole in the aft section of her underside. Instead of *Atlanta*, the damaged *South Dakota* would make the passage to Hawaii with *Saratoga*, along with the cruiser *New Orleans* and five destroyers.

This time, during what proved to be a one-week stay, the crew had the chance to go ashore for some well-earned liberty. Petty Officer McKinney recalled, "The natives were fine-looking people possessed of considerable charm and poise. They were most gracious in their toleration of thousands of our armed forces." McKinney and some of his shipmates were able to borrow bicycles and explore the island. They stumbled on a school to find, to their amusement, the teacher lecturing on the economic geography of New England. Taking a break, the young men feasted on bananas and coconuts. McKinney did note that the local authorities, concerned about the safety of the island's young female population, had placed them under the protection of a barbed wire compound.[8]

Atlanta departed Tonga on Sunday September 13 as Captain Jenkins received orders to intercept a small convoy off Samoa and provide an escort en route to Nouméa. On Tuesday morning, lookouts spotted mastheads on the horizon at the rendezvous spot located 30 miles south of Pago Pago. At that location, the cruiser *Detroit* handed off *Lassen* and *Hammondsport*, and proceeded to Pago Pago with the other ships. Both commissioned ships with Navy crews, the names *Lassen* and *Hammondsport* offered hints of the cargoes contained within. As with *Rainier*, *Lassen* was named for a volcano, and carried some 5,000 tons of ammunition. In contrast, *Hammondsport*, took her name from a small village in the Finger Lakes region of New York that was home to pioneer aviator Glenn Curtiss who sold the Navy its first airplane. Hence, within her hull there were approximately 100 aircraft fuselages, wings, and other aviation-related components.

After four days of steaming, *Atlanta* and her two charges arrived at the Nouméa, New Caledonia. Assigned to a mooring in Dumbéa Bay, those deckhands assigned to the Sea and Anchor detail admired the hilly surroundings but would be disappointed to learn that's all they would see as an epidemic ashore canceled any liberty opportunities. Waiting to come aboard were three new officers, two ensigns from the Naval Academy Class of 1943, which graduated a year early, and a junior grade lieutenant who earned his commission through a reserve program. For the two ensigns, James A. Underwood and Gerald F. Colleran, their trek from Annapolis to catch *Atlanta* placed them in transport which landed Marines at Tulagi on August 7 and on to *Enterprise* on August 24 during the Battle of the Eastern Solomons. In contrast to the young ensigns who drew assignments to the engineering and gunnery departments, Lt. (jg) Frank M. Hereford had been a banker and hoped to serve as a disbursing officer.[9]

Having to settle for eyeball liberty, the crew had little time for shore gazing given the number of chores needing to be accomplished. Late in the morning on Sunday September 20, the tanker *Esso Little Rock* came alongside to provide fuel. After completing that task at 1307, working parties formed to move ammunition. Thanks to nighttime gunnery exercises conducted during the transit, it became quickly apparent that *Atlanta* had been issued a faulty lot of 5-inch star shells. Lieutenant Commander Mustin recalled that of some 45 fired, only two illuminated the night sky. With *Rainier* anchored nearby, 500 alleged flawed shells were removed from the magazines and transported over to the ammunition ship in exchange for illumination rounds that Nickelson and Mustin hoped would work.[10] In addition, to bringing aboard and restocking the magazines, five 600-pound depth charges were added to *Atlanta*'s arsenal to counter the Japanese submarine threat. Those depth charges apparently could have come in handy hundreds of miles to the northeast where Task Force 61 had been steaming en route to Guadalcanal.[11]

Much to the dismay of the officers and enlisted sailors embarked in *Atlanta*, elements of Task Force 61 began arriving at Nouméa to off-load survivors of the carrier *Wasp*. On September 15, the smaller carrier fell victim to Long Lance torpedoes, this time fired from *I-19*. However, unlike *Saratoga*, *Wasp* would not live on to fight another day. Hit by two torpedoes forward on her starboard side, the blasts ruptured gasoline stowage tanks abreast of the forward bomb magazine. With aviation gasoline from the cracked tanks mixing with high-octane fuel gushing from ruptured fueling stations, a combustible vapor quickly ignited into a below waterline inferno that shot up through the forward section of the ship.

From the distance, it appeared as if a volcano had erupted. Additional vapor explosions and the cook-off of bombs stowed in the forward magazine led to a decision to abandon ship. To make matters worse, other torpedoes that had been aimed at *Wasp* found the port bow of the destroyer *O'Brien*. Though the physical damage seemed limited to the unoccupied anchor chain locker, the shock wave that shot back through the hull caused structural stresses that ultimately proved fatal. (Over a month later, en route to San Francisco, *O'Brien*'s bottom gave out. All hands were saved.[12]) However, the spoiled icing on the cake was news that *North Carolina* also took a torpedo hit on her port side abreast of her forward turret. With a gaping hole below the waterline, a second modern battleship was now steaming to Hawaii for repairs.[13]

For the Navy's Theater Commander, Vice Adm. Robert Ghormley who flew his flag on the Nouméa-based submarine tender *Argonne*, the arrival of *Wasp*'s refugees and departure of *North Carolina* limited his options in the greater struggle for Guadalcanal. With only *Hornet* remaining as the U.S. Navy's operational carrier in the region, Ghormley would not risk aggressive actions, leaving the Marines to battle their adversaries for acres of jungle.

Rear Admiral Lee embarked aboard USS *Washington* (BB-56) circa 1942. (Archives Branch, Naval History and Heritage Command, Washington, DC; NH 48283)

Thus, on September 21, *Atlanta* set the Sea and Anchor detail, hauled up and secured her anchor, and got underway, joining with *Helena*, *Salt Lake City*, and three destroyers en route to Task Force 17 centered on *Hornet* and her escorts. The meet up would have personal interest for *Atlanta*'s junior gunnery officer. Lloyd

Mustin's stepfather, Rear Adm. George Murray, commanded the task force that included an additional family element—the destroyer *Mustin*. After a day of steaming during which the three cruisers drilled their gun crews, on the morning of September 23, six ships joined up with Task Force 17. For the first time, *Atlanta*'s crew found themselves in company with sister ships *Juneau* and *San Diego* along with the recently arrived battleship *Washington*. Stationed in a close circular ring around *Hornet*, the four warships alone combined an impressive battery of 34 5-inch 38 twin mounts as well as even larger numbers of smaller caliber AA weapons. Additional AA guns mounted on the cruisers *Helena*, *Salt Lake City*, *San Francisco*, and *Pensacola* would have added additional lethality to the skies over *Hornet*'s flight deck. Meanwhile a screen of destroyers spread out on a wider arc, seeking to engage Japanese submarines before they could position themselves to fire another crippling spread of torpedoes.

The impressive protective shield would not last the day. That evening *San Francisco*, *Salt Lake City*, *Helena*, and four destroyers, detached and headed off in a northwesterly direction from what some in *Atlanta*'s wardroom suspected would be a surface action against the enemy. Then the formation shed additional ships in the morning as *Washington*, *Atlanta*, *Walke*, and *Benham* received orders to proceed to Nukuʻalofa, Tonga. For Lieutenant Commander Mustin, the family reunion had been short. For the newly arrived battleship, this represented a return visit as *Washington* had briefly stopped at Nukuʻalofa to embark Rear Adm. Willis A. "Ching" Lee.[14]

Kentucky-born, Rear Admiral Lee's parents had European roots, so the ethnic nickname came during his years in Annapolis as Lee was a common Chinese surname and he had strong academic interests in the Far East while attending the Naval Academy. He also took keen pride in marksmanship as a member of the Naval Academy's rifle team. Competing at the 1920 summer Olympic Games at Antwerp, Lee captured five gold medals, plus a silver and a bronze.[15]

Lee enjoyed shooting, whether be it '03 Springfields or 16-inch naval rifles, and the voyage to Nukuʻalofa offered an opportunity for the crew of *North Carolina*'s sister ship to impress the rear admiral by expending broadsides at a target on the horizon. *Atlanta* would serve just fine as the target ship, perhaps much to the consternation of the light cruiser's crew, many of whom undoubtably prayed that offsets were correctly plugged into the battleship's fire-control computer. Early on the 25th, *Washington* and *Walke* broke away to a position over 17 miles distant. From *Atlanta*'s pilothouse, those on watch noted that the bulk of the battleship had sunk over the horizon and only the top of her mast remained visible. Lieutenant Commander Mustin, back to the fantail with a measuring device, waited to record the fall of the shot of the first salvo.

With a bright blue sky backdrop. *Atlanta*'s lookouts witnessed a large puff of yellowish-brown muzzle smoke on the horizon. Seconds lapsed. Then astern, three huge splashes appeared where the light cruiser had traversed only a few minutes before. Those topside then heard a crack—the sound of projectiles breaking the sound barrier followed seconds later by the rumble of *Washington*'s guns. As the battleship continued firing salvos, Mustin stood impressed how tight and small the landing patterns were for the salvo groupings. Furthermore, he noted, "they came right down on the wake—they didn't come over or short … meaning that the *Washington*'s battery was beautifully aligned and beautifully calibrated."[16]

Arriving at Tongatabu (or Tonga) mid-morning of September 26, the light cruiser anchored at berth 22. Besides *Washington* and *Walke*, the harbor was empty but for the *Vestal* of Pearl Harbor fame and the troop transport *Barnett*. Berthed alongside *Arizona*, the venerable former collier-turned-repair ship *Vestal* not only survived *Arizona*'s magazine blast, but also two bomb hits and was able to clear the stricken battleship under steam. Since arriving at Tongatabu a month early, the recently repaired *Vestal* had rendered mending services for a wide assortment of visiting damaged warships. *Barnett* sustained damage during the initial landings at Guadalcanal when a Betty bomber crashed into her on August 8.[17]

In the early afternoon, five petty officers mustered with Ensign Straub to go ashore to perform shore patrol duties as Captain Jenkins allowed half of the crew to go ashore on liberty. As before, bicycles were available for touring inland.[18] Horse cart rides were another option for those who chose not to pedal. Unfortunately for one *Atlanta* sailor, a horse pulling his cart suddenly broke into a gallop, throwing him down onto the pavement. With a cracked skull, this crewmember would have to stay behind.[19]

Those who did not have liberty on the 26th could go ashore the following day, a peaceful Sunday. On Monday afternoon, the Sea and Anchor detail was briefly set as *Atlanta* pulled alongside oiler *Sabine* to replenish. Refueling on the fleet oiler's opposite side was the battleship *Washington*. The following morning *Atlanta* shifted to berth 23 and liberty call was once again passed over the 1MC. Ensign Morgan Wesson met some Army nurses and invited the caregivers to experience an evening meal aboard an American man-of-war—an invitation happily accepted.[20]

With Guadalcanal facing no imminent danger, Ghormley apparently had sent Lee with his battleship and escorting light cruiser, and destroyer to Tongatabu to conserve fuel supplies. Thus, *Atlanta*'s sailors enjoyed 12 days in paradise-like surroundings. Morale was further boosted on October 6, with the arrival of mail. Finally, during the early hours of October 7, Captain Jenkins set the Sea and Anchor detail. On the previous day, Ghormley sent an order directing Lee, in command of Task Group 17.8, to head west in *Washington* with *Atlanta* and the destroyers *Benham* and *Walke*. For Jenkins, he now headed into battle with a more senior wardroom—at least on paper. A new ALNAV emerged from the radio room, announcing the promotion seven of the light cruiser's junior officers.[21]

Over the next 72 hours, the four ships pushed westward, entering the Coral Sea late on October 9. Concerned about the torpedo threat, Jenkins kept his ship at Condition 3, which kept watch standers at ship control, damage control, and engineering stations, as well as an assortment of gun mounts. The three Condition 3 officers of the deck included the Communications Officer Lt. Cdr. Paul Smith, Assistant Engineer Lt. Cdr. Jack Wulff, and finally Assistant Gunnery Officer Lt. Cdr. Lloyd Mustin. To protect his capital ship from enemy underwater missiles, Lee positioned *Atlanta* and the two destroyers to form an anti-submarine screen. Unfortunately, the 18 knots average speed rendered sonar equipment ineffective. However, speed combined with an effective zigzag plan that could fend off a torpedo targeting solution, would prove to be an adequate trade-off.

CHAPTER 10

Action Over the Horizon

The journey of *Atlanta* as part of Task Group 17.8 in company with the battleship *Washington*, and two destroyers *Benham* and *Walke* proved relatively uneventful. On the late afternoon of October 9, sonar reported a contact. The Officer of the Deck, Lieutenant Commander Mustin, sent the crew to General Quarters and ordered the helmsman to steer the light cruiser towards the bearing of the sound. However, after a few minutes, sonar lost the contact and Mustin had the helmsman return *Atlanta* back to the original heading.

During the following morning, the last of independent steaming for Task Group 17.8, Jenkins's ship refueled from *Washington*'s significantly larger bunkers. Then in the early afternoon, one of *Washington*'s scout planes pulling a target sleeve provided *Atlanta*'s crew with an additional target practice opportunity.

On the morning of October 11, the four ships came upon Task Group 62.6, centered on *Zeilin* and *McCawley*, located some 200 miles north of the northwest end of New Caledonia. Having departed from Nouméa, with a regiment of 2,837 Army soldiers along with 210 ground crewmen from the 1st Marine Air Wing, the two transports also carried jeeps, trucks, heavy guns, and 4,200 tons of cargo. *Washington*, *Atlanta*, *Benham*, and *Walke* fell in with the convoy to provide added protection to assure these critically needed reinforcements arrived at Guadalcanal. Following radio reports on the situation on "Cactus," Jenkins and his senior officers appreciated the importance of getting the two transports to their destination.[1]

As the small convoy forged ahead, Japanese air raids prevented the polyglot of Navy, Marine Corps and Army aircraft collectively known as the "Cactus Air Force" from lashing out at an incoming Japanese bombardment force which would vanguard a small flotilla of Japanese reinforcements. *Atlanta*'s radiomen monitored sighting reports of three Japanese cruisers and six destroyers coming down the Slot. The Japanese force, commanded by Rear Adm. Aritomo Goto, consisted of three cruisers and two destroyers that trailed a reinforcement convoy centered on seaplane carriers *Nisshin* and *Chitose*, and six destroyers. As the eight warships dropped off armaments, supplies, food, and fresh troops, Goto's five ships covering the supply effort would surge forward to shell Henderson Field.[2]

Unfortunately, Guadalcanal remained too far distant for *Atlanta*'s, and more significantly, *Washington*'s guns to come to defend the Marines ashore, however, another American surface task force was in the vicinity. Task Force 64, commanded by Rear Adm. Norman Scott, featured four cruisers and five destroyers. Following the message traffic, Jenkins and his officers knew that Vice Admiral Ghormley had directed Scott to attack the Japanese "if conditions were favorable"—guidance that Mustin winced at as "weaseling." Scott, however, was not prone to weasel. On *Atlanta, Washington*, and the other ships of the small convoy steaming northward through the night, watch standers paced back and forth awaiting updates from the sea battle taking place off Cape Esperance off the northwest tip of Guadalcanal.

To meet Gato's force, Scott steamed in a column led by the destroyers *Farenholt*, *Duncan*, and *Laffey* leading the cruisers *San Francisco*, *Boise*, *Salt Lake City*, and *Helena*, which were trailed by the destroyers *Buchanan* and *McCalla.* Scott was aware of the Tokyo Express group that had arrived off Guadalcanal but understood he needed to engage the bombardment group.[3] Thus, he placed his column to form a wall between the resupply group and the oncoming Japanese cruisers. Steaming in a northeasterly direction, there came a point where Scott would have to reverse course to maintain a broadside to an oncoming enemy force. At 2333, Scott ordered his column to port so as to reverse course to steam in a south-westerly direction. Such a "column" maneuver called on *Farenholt* to swing around to the left with *Duncan* and the remaining ships following the lead destroyer's wake. In retrospect, had the maneuver been properly executed, it could have had the lead destroyers crossing in front of the 8 and 6-inch guns of Scott's cruisers as American radar sets unveiled the presence of the oncoming foe. However, the bridge watch team in *San Francisco*—Scott's flagship—misconstrued the radio signal transmitted from Scott's perch one deck below and the conning officer ordered an immediate turn to port. *Boise* and the remaining ships followed in *San Francisco*'s wake. Of the three destroyers that had been in the van, *Farenholt* and *Laffey* eventually caught up with the reversed column, unfortunately positioning themselves between the American cruisers and Japanese surface force. Detecting the Japanese on radar, *Duncan*'s commanding officer, Lt. Cdr. Edmund B. Taylor took the initiative to flank the oncoming enemy force to launch a torpedo spread.[4]

The opposing admiral approached in a "T" formation with Goto's cruiser *Aoba* at the head of the "T" and destroyers *Fubuki* and *Hatsuyuki* off the flagship's starboard and port beams. Cruisers *Furutaka* and *Kinugasa* trailed *Aoba*. Though his lookouts spotted Scott's ships crossing ahead and identified them as enemy, Goto believed they were looking at the resupply group and ordered recognition signals sent. *Helena* answered the friend or foe query with a broadside. Shortly thereafter, shells pierced *Aoba*'s flag bridge killing many watch standers and inflicting wounds on Goto that would prove fatal. American gunners also scored hits on *Aoba*'s forward turrets and gun director as well as on the following cruiser, *Furutaka*, where a shell which hit on a torpedo tube set off a secondary explosion that exposed the ship to further salvos. The skipper of *Kinugasa* prudently turned away along with the destroyer *Hatsuyuki*—though the latter sustained two hits.

Despite Goto's blunder, the Americans also suffered casualties. Positioned between the oncoming Japanese and Scott's cruisers, *Farenholt* was hit by an American 6-inch shell just above the waterline and bursting shell fragments killed a number of the topside gun crews and lookouts. The rogue *Duncan* fired two torpedoes at Gato's force before being struck by shells from both engaging navies. Furthermore, an 8-inch shell from the unscathed *Kinugasa* penetrated *Boise*'s forward barbette, setting off gunpowder and stowed ammunition that proved fatal for 107 sailors. Fortunately, buckled hull plating below the waterline allowed seawater to douse the inferno, protecting the forward black powder storage hold. *Boise* would live to fight another day.[5]

Salt Lake City received some light blows from *Kinugasa* but also inflicted wounds on the Japanese cruiser. Eventually *Kinugasa* and *Hatsuyuki* would escape to the northwest, escorting a crippled *Aoba* which suffered at least 40 hits. The cruiser *Furutaka* would not make it back—going down before sunrise. The sea would also claim the destroyer *Fubuki*.[6]

Savaged by gunfire from both warring navies, *Duncan*'s survivors abandoned the Federal Shipbuilding and Dry Dock Company built ship in the early hours, fearing flames would reach the destroyer's magazines with devastating consequences. Fortunately for these sailors, another Kearny-built destroyer—*McCalla*—came along to investigate the blazing hulk. After picking up survivors, any hopes of salvaging the charred destroyer—which had stayed afloat—were dashed by a bulkhead failure that led the ship to sink in the late afternoon.[7]

The outcome of the night engagement proved a morale boost to those serving in *Atlanta*, though victory could have been more total had Scott turned on the resupply group which accomplished its mission of getting reinforcements ashore. However, since Scott interdicted Goto's force, the Cactus Air Force—spared of an overnight shower of steel—lifted at sunrise to conduct torpedo and bombing attacks on the retreating Japanese forces, successfully sinking two more Japanese tin cans.[8]

As the Japanese struggled to depart to the northwest, the American counter-reinforcement centered on *Zeilin* and *McCawley* proceeded along waters up the eastern side of Guadalcanal. At 1438, the formation of ships turned westward and increased speed to 18 knots to traverse the waters along Guadalcanal's northern shore, aiming to arrive at Lunga Point before dawn minus *Washington*, *Atlanta*, *Benham*, and *Walke*. Believing the waters off Lunga Point to be restrictive for battleship maneuverability, the Americans reversed course for those four warships at 1650. Over an hour later, those on topside watch on *Atlanta* witnessed the sun set over Guadalcanal to the west. Lee's group would remain in a holding pattern all day on the 13th as *Zeilin* and *McCawley*—guarded by the destroyers *Gwin, Nicholas,* and *Sterett*—offloaded their precious human cargo. It might be surmised that the decision to keep the battleship, light cruiser, and two destroyers out of harm's way may have been questioned as reports of two Japanese air raids filtered back. Lumbering Betty bombers would have made for inviting targets for *Atlanta*'s eight 5-inch 38 twin mounts. Fortunately, the Japanese airmen focused on the American airfields ashore, sparing the two valuable troop transports. Though the Japanese bombs would destroy a parked B-17 and damage others, replacement aircraft could be flown in. Sinking troop transports with troops still embarked could have been more of a game changer. That evening, with their mission nearly complete, the Japanese finally turned their attention towards *Zeilin* and *McCawley*. Using a 150mm artillery piece landed on the previous night, the Japanese targeted the ships off Lunga Point. After a shell splashed an uncomfortably close 25 yards off *McCawley*, the three guarding destroyers charged westward towards the Point Cruz enemy positions to attempt to silence the offending big gun.[9]

Having landed the troops ashore, the two transports and their tin can escorts retraced their path back to the eastern side of Guadalcanal where Lee's four ships waited. With *Washington* and her nine 16-inch rifles and *Atlanta*, and two destroyers with additional 5-inch guns in close proximity, one of the "What if's" of World War II was "What if" Lee's force had been positioned off Lunga Point that evening to intercept the Japanese battleships *Kongo* and *Haruna* sent by Admiral Yamamoto to take Henderson Field out of commission long enough to allow slower Japanese transports to remain off Guadalcanal in daylight without fear of air attack? Somehow the approach of the two battleships screened by a light cruiser and nine destroyers evaded Allied scout planes. Had Ghormley been alerted to their presence, he would have had the option of deploying the more modern *Washington* with her heavier caliber guns against the two World War I-era modernized dreadnoughts. *Atlanta* would have experienced her first surface action. Instead, when *Kongo* and *Haruna* opened on the American perimeter on Guadalcanal, the only response came from 5-inch coastal batteries that lacked the range to counter the incoming bombardment. Eventually, four torpedo boats scrambled to launch torpedo attacks that failed to land blows. Thus, as

Atlanta's non-watch standers slept comfortably in their bunks as the cruiser headed south between Guadalcanal and San Cristobal, the Marines and newly arrived Army troops endured the arrival of 973 14-inch shells. After expending specially designed fragment rounds that scattered shards of steel over wide radii, the two battleships cratered the landscape along Henderson's runways and adjacent tarmac with armor-piercing rounds. For the poor American souls on the ground that night, the impact of the shells rocked the earth, creating a sense of helplessness that would never be forgotten. Thankfully, because the airfields bore the brunt of the incoming ordnance, the massive bombardment claimed only 41 of the estimated 20,000 Americans ashore that evening.[10]

As the first rays of light began to penetrate the jungle canopy, Japanese observers on the hills overlooking the American perimeter were stunned to see the takeoff of the morning scout mission. Though Henderson Field was rendered unusable, a satellite fighter strip with 30 aircraft remained operational. This would not bode well for Japan, given the objective to totally eliminate American land-based air power. Spotted by the American Dauntlesses some 140 miles distant, the convoy needed to be stopped—but by what and whom? Making matters worse, a noontime raid on Guadalcanal by 26 Betty bombers could not be met by the ground-based fighters due to limited fuel supplies. A second attack was resisted. Both attacks interfered with efforts to assemble a limited attack on the oncoming *Marus*. Finally, the Americans launched a desperation strike in the late afternoon as the Japanese convoy began its final approach, scoring only a light hit on one of the escorting destroyers.[11]

On the bridge of *Atlanta*, Jenkins followed the message traffic knowing that *Atlanta* as well as *Washington* retained enough fuel stocks to remain in the vicinity of Cactus for at least another day. Meanwhile *Benham* detected a submarine contact. Pursued by both *Benham* and *Sterett*, the enemy submarine evaded the surface threat. At 1330, lookouts spotted a Japanese scout plane on the horizon. With *Atlanta* at General Quarters and tracking the observer, the Japanese pilot wisely kept his distance and reported the sighting of the departing ships—a report that added confidence about the potential success of the pending reinforcement effort.[12]

The Japanese reconnaissance plane came on the Americans just after Task Group 17.8 parted company with the pair of transports and their destroyer escorts heading back to Nouméa. Lee's four warships trudged on to Espiritu Santo where they would refuel and then meet with Scott's flotilla. As *Atlanta*'s deck watches executed the zigzag plan during the night, several thousand Japanese troops began to scamper ashore as the six *Marus* arrived around midnight which must have been demoralizing for the Marine and now Army defenders surrounding Henderson Field.[13] Adding to the misery, two Japanese cruisers took their turn at raking Henderson Field with shellfire. Though not as severe as the night before, the 752 8-inch shells added additional craters and ripped apart aircraft that were being patched up from the previous night's pounding.

Despite the overnight bombardment, the Cactus Air Force still had a few aircraft that it could employ against the six *Marus* which needed to spend the day close ashore to complete off-loading troops and supplies. At first, six Wildcats broke through the air cover provided by the aircraft carriers *Hiyo* and *Junyo* to strafe the enemy merchantmen. By mid-morning, the Cactus Air Force launched a credible strike of two dozen dive-bombers and fighters as well as Catalina that managed to launch a torpedo that punched a hole into the *Sasako Maru* that proved fatal. The Army Air Forces also took aim at the static targets as a flight of B-17s from Espiritu Santo caught the *Azumasan Maru* with a noontime bomb. Later that day, the Cactus Air Force claimed *Kyushu Maru*.[14]

Given the location of the airfield near Luganville at Espiritu Santo which had been expanded and reinforced over the previous months by Seabees to handle the weight of Army bombers and other military aircraft, *Atlanta*'s topside sailors likely noted the departure of the Guadalcanal-bound flight of B-17s in the morning and their return later that afternoon. Jenkins's ship had arrived late the previous day, steaming around the southern portion of the island to reach a strait that led to Luganville. In the months since the attack on Pearl Harbor, the tropical paradise that had been a joint colonial outpost of both London and Paris had been transformed into a major American logistical facility which would later serve to inspire Richard Rodgers and Oscar Hammerstein's *South Pacific*. Pulling alongside the anchored oiler *Kankakee*, *Atlanta* had her bunkers topped off. Across the way, crewmembers stepping out for fresh air could see *Boise*, noting the forward two burned turrets. As the light cruiser replenished, new directives had arrived over the wireless. As of 1800, a reorganized Task Force 64 built around the battleship *Washington* came into being. Besides incorporating the other three ships of Task Group 17.8, the new command to be led by Rear Admiral Lee would absorb the remnants of Scott's force to include the cruisers *San Francisco*, *Chester*, and *Helena*, and the destroyers *Aaron Ward*, *Buchanan*, *Lansdowne*, *Lardner*, *Laffey*, and *McCalla*.[15]

Once again Captain Jenkins set the Sea and Anchor detail for an evening departure. Joining up with *Washington*, *Walke* and *Benham* at 2200, the four ships sped up to 22 knots on a northwesterly heading for a morning rendezvous. Following the merger of the two forces, Lee steamed his new task force towards a position southwest of Guadalcanal from where they could surge up to intercept further Japanese resupply efforts. A Japanese Kawanishi 97 four-engine floatplane appeared on the horizon in the late morning to shadow the powerful American surface group. Anticipating that reports back to Rabaul would instigate air attacks, Captain Jenkins sent his crew to General Quarters at 1215. Fortunately, the weather provided a reprieve. Haze limiting visibility to four miles and a cloud ceiling below 3,000 feet cloaked the task force from airborne observers. Jenkins deciding to stand down at 1239, though he kept the gun mounts manned. At 1400, two enemy aircraft detected by *Atlanta*'s SC radar flew by oblivious to what floated below.

Unfortunately, the decision to send *Washington* and *Atlanta* to top off their bunkers when they could have rendezvoused with Scott's force a day earlier meant that Task Force 64 remained too distant from Cactus to intercept the three departing *Marus* and their escorts. Apparently, it was later discerned that the refueling directive was an administrative mistake in that Ghormley's logisticians had ordered Task Force 17, centered on *Hornet*, to refuel, forgetting that Lee's Task Group 17.8 was a component of this force—at least on paper. The intent all along had been for Lee to join up with Scott a day earlier without topping off. Unfortunately, intent meant that those ashore on Cactus would have to endure yet another night of bombardment as two heavy cruisers escorted by destroyers, unleashed 926 8-inch shells and 253 5-inch rounds at the sleep-deprived Americans.[16]

Instead, at 1800, Task Force 64 reversed course to a southeasterly heading to stay out a range of airborne snoopers, and would halt at a point 200 miles south of Rennell Island, a coral atoll due south of Guadalcanal. However, two destroyers forged on to Cactus that evening. The destroyers *Aaron Ward* and *Lardner* arrived early the next morning to awaken the 5,000 recently arrived Japanese troops with some 5-inch ordnance.[17]

The gunfire from the two American tin cans was but the opening salvos of the first full day ashore for the new arrivals. Henderson Field remained operational, and the Cactus Air Force launched 58 combat sorties against those who had gotten off the six *Manus* the previous day. Making matters even worse for the new arrivals, they also had to seek cover to evade attacks from *Hornet*. Ghormley finally allowed his lone carrier to surge forward from New Caledonia to join attacks on the Japanese positions at Guadalcanal, as well as Japanese seaplane facilities up the Slot at Rekata Bay on the northern tip of Santa Isabel.[18]

The assertive decision to bring *Hornet* back into the fray proved too little too late for Ghormley. Later on October 16, Admiral Nimitz requested and received permission to sack his South Pacific commander and replace him with Vice Admiral William F. Halsey.[19] Two days later, when Halsey arrived at Nouméa in advance of Task Force 16, centered on the repaired *Enterprise*, he was presented with a sealed envelope to learn he was to be fleeted up from his anticipated role as a task force commander.

For the wardroom of *Atlanta*, the news of the change of command injected optimism. Lieutenant Commander Mustin scribed in his diary: "There'll be some changes made in the Solomons, Mr. Moto—this is the greatest shot in the arm we've had this war." Having arrived in Hawaii in the wake of the Doolittle Raid months earlier, few doubted Halsey's willingness to act aggressively. Halsey would fulfill expectations. Unlike his predecessor, Halsey did not hesitate to fly to Guadalcanal to confer with Major General Vandegrift who expressed confidence that his Marines and Army troops could hold against an anticipated Japanese ground assault with the caveat that the Navy needed to provide more active support. Halsey assured him he would have that support. Likewise, Admiral Yamamoto was making a similar commitment to Vandegrift's counterpart by assembling the Combined Fleet for a thrust into the Solomons with an objective to clear American naval forces from the region as Japanese soldiers reclaimed Henderson Field.[20]

The Japanese Ground Commander, Lt. Gen. Masao Maruyama had anticipated overwhelming the Americans through a flanking maneuver around the Marines who had dug in from the northern coast along the western approaches to Henderson Field. A feint against these positions supported by tanks and artillery would draw the bulk of Vandegrift's forces while the bulk of Maruyama's troops would emerge from the jungles along the lightly defended perimeter south of Henderson. Unfortunately for Maruyama, that jungle proved to be an American ally. For the thousands of Japanese soldiers who had to traverse the rugged landscape and hack through the underbrush, it proved impossible to reach the attack launch points planned for the evening of the October 22. Loaded down with arms, ammunition, rations, and water, the exhausted enemy troops finally arrived at their jump-off points two days later, opposite of Marine defenders commanded by Lt. Col. Lewis B. "Chesty" Puller.[21]

Though Halsey had committed support to Vandegrift, that support would not come in the form of Task Force 64. Lee's surface force continued to steam in a holding pattern west of Santo Espiritu, some 350 miles south of Cactus. On the morning of October 19, upon receiving intelligence about Japanese warships heading towards Cactus that evening, Lee ordered his ships to accelerate to 20 knots on an interception course of 335 degrees. However, at 20 knots it would take nearly 18 hours to reach Guadalcanal. Lee, upon realizing he had too much ocean to steam over, slowed the speed of the formation to conserve fuel. Instead, orders came down to divide the large gaggle of surface ships into two smaller task groups. Task Group 64.1 would include *Washington, Atlanta,* and *San Francisco. Helena* and *Chester* provided the bulk of the firepower for Task Group 64.2. The escorting destroyers were split between the two groups. For the remainder of the day, the two task groups stayed within sight of each other.[22]

On the next day October 20, *San Francisco* joined with *Helena*, *Chester*, and four destroyers. Later that evening those seven ships were heading back towards Espiritu Santo within sight of Task Group 64.2 also steaming on a southeasterly heading. At 2120, *Atlanta*'s war diary recorded: "Sighted explosion and gunfire to southward in the direction of Task Group 64.2 ..." *Atlanta*'s lookouts had witnessed a torpedo detonation that had been launched by a Japanese submarine against *Chester*. On the light cruiser, many of the non-watch standers had already crawled into their bunks following the evening GQ ritual when the call to battle stations alarm sounded.[23]

Captain Jenkins had *Atlanta* follow *Washington*'s movements as everyone monitored the radio chatter to piece together the tactical situation. At 2158, *Atlanta*'s war diary recorded the blow *Chester* received, and Lee sent over *Walke* and *Aaron Ward* to reinforce the screen surrounding the crippled cruiser that had absorbed a hit amidships on the starboard side.[24] *Chester*'s damage control teams contained the flooding to the forward engine room and number 3 fireroom. Could the cruiser evade further attack to make it back to Espiritu Santo as the first leg of a long journey back to the States for extensive repairs? The next day *Washington* and *Atlanta* caught up with the crippled warship to provide protective cover in the advent of air attack—a possibility given the presence of a Kawanishi 97 on the horizon. Fortunately, deteriorating weather conditions mitigated chances for enemy air action. *Atlanta* steamed slowly alongside the damaged cruiser that struggled to maintain 5 knots, given that one of her four shafts had snapped. Mustin judged her to be in bad shape, surmising if *Chester* attempted to accelerate, that shaft could be pulled out of the hull.[25]

While the other cruisers surged ahead to Espiritu Santo to refuel and take on provisions, *Atlanta*, *Washington*, and two destroyers protected the vulnerable heavy cruiser on her prolonged voyage south. Following the slow trek upon return to Espiritu Santo on the morning of October 23, Captain Jenkins set the Sea and Anchor detail as *Atlanta* steamed through the Bougainville Straits along the southeastern side of Espiritu Santo to eventually anchor in Segond Channel. *Chester* followed astern. Though hardly comparable to *Chester*, *Atlanta* had to confront some propulsion challenges during the recent sortie. While anchored, Lieutenant Commander Loeser's "snipes" worked to patch leaky tubes in the water wall on the superheater side of boiler #3. Unfortunately, for Loeser, the long-term prognosis for boiler #3 was not good, unless new tubes could be installed during a shipyard availability. At best, the boiler could be operational for only a few more hours. Captain Jenkins would need to depend on steam generated from boilers 1, 2, and 4. Also during *Atlanta*'s short turnaround, she received some 200,000 gallons of black oil courtesy of the tanker *Willamette*. More welcome than the black oil was weeks' backlog of letters from the States. In the early afternoon, Jenkins again set the Sea and Anchor detail. As the light cruiser's anchor lifted off the channel bed, the word "Underway Shift Colors" was passed. *Atlanta*'s 1548 war diary entry recorded: "Underway in company Task Group 64.1 and stood out to sea. When clear of minefields formed circular screen on WASHINGTON. ATLANTA in station 4000. LANSDOWNE, LARDNER, McCALLA, AARON WARD and BENHAM on circle 4 also base course 270 degrees (T) at 18 knots."[26]

The next day, October 24, Task Group 64.1 steamed well west of Espiritu Santo, and Task Group 64.2 with *San Francisco* and *Helena*, and their destroyer screen steamed just on the horizon. Once, a hovering big four-engine seaplane reported back to Admiral Yamamoto with the position of the two surface groups.[27]

That evening, the Japanese infantry thrust towards American lines south of Henderson Field. Lieutenant Colonel Puller, with the help of the National Guardsmen and artillery, fought off the Japanese onslaught. Confident of the outcome of Maruyama's attack, the next morning a Japanese reconnaissance plane buzzed over Henderson Field, intent on confirming the win. Marine AA gunners brought it down. A similar greeting was given to Japanese destroyers that were sent to polish off remnants of Vandegrift's Leathernecks who were expected to be clinging on the beach hoping for evacuation. They would be driven away.[28]

The Japanese commanding general rallied his troops to attempt one more assault after sunset. The Americans made effective use of canister shot and well-positioned machine guns to cut down hundreds of Imperial soldiers. Others were killed in hand-to-hand combat. With his offensive capability sapped by the loss of over 2,000 troops, Maruyama and his survivors had little choice but to backtrack.[29]

Had Maruyama been able to coordinate with those Japanese Navy surface units—which had appeared earlier in the day—to stick around to lay a barrage on Puller's positions, the outcome of the battle may have been different. However, the Japanese surface forces chose to flee Ironbottom Sound. Had they stayed though, they probably would have been annihilated by the powerful American surface force coming up from the south. That Sunday morning, Halsey ordered Lee to go after the Japanese combatants that were off Lunga Point exchanging gunfire with Marine batteries ashore, and two former World War I-vintage flush deck destroyers that had been converted to minesweepers. Lee's task force, located southeast of Rennell Island, had some distance to cover to make it up to Cactus that evening, even with *Washington* capable of 26 knots. Conceptually, they would arrive around midnight to block the approaches leading from Ironbottom Sound to hopefully engage any remaining Japanese surface forces. At 1338, *Atlanta*'s lookouts sighted Rennell to the northwest which meant that any Japanese observers placed on the island would be seeing a large formation of ships to the southeast! Though there was low visibility as well as a low ceiling, Captain Jenkins took the precaution at 1400 to call the crew to General Quarters. As formation passed to the east of the coral atoll, Lee turned due north to begin the final leg of the transit, increasing speed to 26 knots at 1536. At 1600, the formation came left to a northwesterly heading. At 1743, with the mountaintops of Guadalcanal visible on northeastern horizon, *Atlanta*'s war diary reported: "Took battle formation, ATLANTA in support of van destroyers, heavy ships in column astern. Changed course to 323 degrees (T)."[30]

With 10 destroyers at his disposal, Lee placed six of them in the van, supported by Jenkins's ship. *San Francisco*, *Helena*, and *Washington* followed, forming the main body. Four additional destroyers brought up the rear. Based on previous engagements where rain squalls hid opposing forces until they were almost on each other, Mustin anticipated a close-range "shotguns-across-the-dinner-table sort of affair." However, the distant mountaintops of Guadalcanal visible at 50 miles indicated weather conditions had dramatically improved. With a near full moon any radar advantages held by the Americans could be negated. By 2000, following sunset, Task Force 64 was well to the southwest of Guadalcanal. At about 2100, *Washington*, *San Francisco*, and *Helena*, launched floatplanes. The 2154 entry in *Atlanta*'s war diary read: "Changed Couse to 000 degrees (T)." Passing west of the Russells, a pair of small volcanic islands approximately 25 miles northwest of Cape Esperance, Lee slowed the formation speed to 22 knots.[31]

Lee then wheeled the formation to the northeast, then east, and then to the right once more to head directly at Savo with the hope of intercepting any Japanese combatants deployed to support Maruyama's last-gasp effort. Perhaps both to Maruyama's and to Lee's disappointment—none were to be found. As previously noted, the Japanese surfaces forces had departed before sunset. The empty waters around Guadalcanal were confirmed by three aircraft that had been launched to seek out any enemy warships in the vicinity. Throughout the evening, *Atlanta*'s guns crews, damage control teams, and others stood ready at their GQ stations anticipating the exchange of shotgun blows. At 0230, Captain Jenkins instructed his officer of the deck to secure from General Quarters.[32]

Convinced that no surface battle would occur that evening, at 0215 early on the 26th, *Atlanta*'s crew left their battle stations and Condition 3 watches were set. At 0315, now heading due south, Task Force 64 formed a column formation with the major combatants following *Atlanta*'s lead while the destroyers steamed parallel to Jenkins's light cruiser in a line abreast. At 0535, Task Force 64 once again divided with Task Group 64.1, centered on *Washington* and *Atlanta* continuing on south. At 0605, *Atlanta*'s lookouts spotted a Japanese seaplane, "well beyond gun range." By 0800, Task Group 64.1, now well clear of Cactus, steamed in waters west of Rennell Island."[33]

With the Japanese having suffered a serious defeat on the ground, Lee's task force retired to the south at 25 knots. Again Japanese naval aviation kept Lee under surveillance, initially with a Nakajima single-prop floatplane and then with a much larger Kawanishi 97. Expecting air attack, Jenkins kept his crew at battle stations. As *Atlanta*'s crew again passed empty time at their stations, other American sailors at GQ stations several hundred miles over the northeastern horizon were fighting for their lives. On the bridge, Jenkins followed reports emerging from waters northeast of Guadalcanal in the vicinity of the islands of Santa Cruz which would lend their name to this engagement.

The return of Task Force 16 commanded by Rear Admiral Thomas Kinkaid on October 24 not only doubled American carrier strength with *Enterprise* returning to the order of battle, but also doubled the number of battleships in theater as *South Dakota* steamed with *Big E.* Kinkaid also had the heavy cruiser *Portland*, light cruiser *San Juan*, and eight destroyers.[34] Kinkaid's force hardly had time to settle as new management in Nouméa directed Task Force 16 to rendezvous with George Murray's Task Force 17 northeast of the New Hebrides at midday on October 24 to form Task Force 61 under the more senior Kinkaid.[35] Of note, within the new formation steamed antiaircraft protection in the form of *Atlanta*'s three sisters—the aforementioned *San Juan* as well as *San Diego* and *Juneau*.

Task Force 61 came together just as Admiral Yamamoto surged his Combined Fleet south to cover the presumed Japanese seizure of Henderson Field, having an added objective of engaging enemy naval forces for that decisive naval battle that would avenge Midway and weaken American resolve to continue the war effort.

Yamamoto split his forces; Vice Adm. Nobutake Kondo who served as the overall Japanese on-scene commander, flew his flag with the support force that included the carrier *Junyo* escorted by two battleships, five cruisers, and 10 destroyers. A main body followed, with Vice Admiral Nagumo's carriers *Zuikaku* and *Shokaku*, and light carrier *Zuiho* guarded by a cruiser and eight destroyers. Similar to what Halsey had with Lee's Task Force 64, Yamamoto formed a Vanguard Force commanded by Rear Adm. Hiroaki Abe that featured two battleships, four cruisers, and seven destroyers.[36]

The Japanese advantage in carriers and surface combatants did not deter Halsey. As with the Japanese, the American inventory of long-range aircraft including PBYs and B-17s, scoured the Southwest Pacific for enemy vessels and fed back sightings of the approaching Combined Fleet throughout midday on the 25th. Kinkaid, hoping to get in a first blow, launched a strike from *Enterprise* to go after Nagumo's flattops. However, the veteran of the Midway debacle, knowing he had been sighted, wisely reversed course to elude the oncoming attackers.[37]

The bright moon that made it easy for Lee's three scout planes to discern there were no enemy warships remaining in the vicinity of Guadalcanal also illuminated the Combined Fleet for night-flying Catalinas equipped with airborne radar as well as some ordnance. Having again turned on a southerly heading to close on Guadalcanal and any defending American naval forces, Nagumo received a rude surprise at 0250 as one of four bombs dropped by one of the Catalinas detonated a mere 300 yards off *Zuikaku*. The cautious Nagumo again changed course to a northerly heading. Kinkaid, having sent out an airstrike the previous afternoon with no results, dispatched and fanned out eight pairs of Dauntlesses from *Enterprise* to confirm the overnight Catalina reports.[38]

One of these pairs not only sighted Nagumo's carriers but dove on *Zuiho*, scoring a hit aft that put that floating airfield out of service.[39] Unfortunately for Kinkaid, most of *Zuiho*'s aircraft were already airborne along with those of Nagumo's larger two carriers as Task Force 16 had been discovered at first light by a Japanese scout plane. As *Zuiho*'s damage control teams extinguished blazes back on the stern, two waves, totaling 110 aircraft, winged their way towards Task Force 61.[40]

Reacting to the sighting reports from the *Enterprise* scouts, Kinkaid responded, and the Americans sent 64 aircraft in the direction of Nagumo's carriers. In this naval shootout the Americans were late to the draw and would not be able to put enough bullets on target. Some of those bullets would be lost en route as the two masses of aircraft crossed paths and the Japanese thinned out the oncoming American herd shooting down eight aircraft, losing four of their own.

With Task Force 16 cloaked by a rain squall, the first wave of the Japanese air armada focused on Task Force 17—more specifically *Hornet*. At 0910, Japanese dive-bombers who had evaded the American combat air patrol began their individual dives on the carrier that had launched the Doolittle raid six months earlier. Despite evasive maneuvers and ferocious AA fire that brought down many of the attackers, within minutes, the Japanese airmen scored three bomb hits and caused additional damage when a Val crashed against *Hornet*'s stack and penetrated down to the crew's galley. As dive-bombers descended on *Hornet*, torpedo planes approached the carrier from separate vectors, scoring a hit on the port side into the forward engineering space while the other slammed into it on the starboard quarter into compartmentalized spaces. Rear Admiral Murray's flagship would suffer one more blow as a burning Val plunged into the port side of the flight deck, eliminating several gun mounts and killing their crews and starting yet another blaze below in the hangar.[41]

As the surviving Japanese attackers regrouped to return to their carriers, *Hornet*'s dive-bombers scored hits on *Shokaku*, tearing into her flight

A Japanese Type 99 shipboard bomber (Allied codename Val) trails smoke as it dives toward *Hornet* (CV-8), during the morning of 26 October, 1942. This plane struck the ship's stack and then her flight deck. A Type 97 shipboard attack plane (Kate) is flying over *Hornet* after dropping its torpedo, and another Val is off her bow. *Hornet* was lost in the battle. (Archives Branch, Naval History and Heritage Command, Washington, DC; 80-G-33947)

deck. A second group from *Hornet* located and delivered a savage attack on the cruiser *Chikuma*.[42]

Meanwhile, Nagumo's second wave, aware of the presence of *Enterprise* as one of the attackers of the exiting first wave, caught a glimpse of Task Force 16 emerging from the squall, and ignored the burning *Hornet* and successfully located Kinkaid's flagship. As with *Hornet*, the Japanese pilots executed their attack with skill and bravery and scored two hits and a damaging near miss at great cost. Not only did the diving aircraft have to avoid the rising ordnance from the *Big E*'s gun crews, but also fusillades of steel fired from the nearby *Portland*, *San Juan*, and battleship *South Dakota*.[43]

In contrast to the well-timed simultaneous torpedo plane attack on *Hornet*, the Japanese came in late, allowing the American gunners to savage the incoming low fliers. Unable to complete his run on *Enterprise*, a pilot of a damaged Kate chose instead to plow into a 5-inch mount of the destroyer *Smith*. Of the 110 aircraft that Nagumo had launched against the two American carriers, 49 would not return. However, the sacrifices made had tipped the tactical advantage greatly in favor of Japan. Though Kinkaid's planes had taken *Zuiho* and *Shokaku* off the playing field, *Zuikaku* and *Junyo* remained unscathed, and the Japanese held an advantage in surface forces. Indeed, a wave of aircraft off the latter ship arrived over Task Force 16 at 1121 and pressed a dive-bomber attack during a rain squall that forced the Japanese dive-bomber pilots to approach on a lower guide slope. One near miss to starboard shuddered the ship and caused damage below *Enterprise*'s waterline. Meanwhile, *Atlanta*'s sister *San Juan* became the first of the class to suffer a blow as a bomb passed through her stern, exploding and jamming the rudder in a right full position. A bomb also landed on the forward turret of *South Dakota*. While the turret survived unscathed, bomb fragments killed two and wounded some fifty other topside bluejackets.[44]

As *Junyo*'s planes departed, Murray shifted his flag to *Northampton* which attempted to take the crippled *Hornet* under a tow, and *Enterprise* successfully trapped 57 aircraft that had flown off the two carriers earlier that morning. Unfortunately, a pilot of one of the returning torpedo planes decided to ditch ahead of the destroyer *Porter* and in doing so unintentionally launched his torpedo at his intended rescuer. It struck the destroyer between boiler rooms, killing 15. Deciding the damaged tin can was expendable, Kinkaid directed another destroyer to pull off *Porter*'s surviving crew and sink her with gunfire.

At this point Kinkaid faced an even more difficult decision. Should Task Force 16—centered on a limited operational *Enterprise*—remain in the vicinity to cover an attempted recovery of the more crippled *Hornet* and risk decimation from additional Japanese air strikes and potential surface action with Kondo's battleships, or should he spare the remaining carrier in the South Pacific from potential harm and depart to leave *Hornet* to fend for itself. He chose the latter option, sealing the fate of a carrier whose crew had contained the fires and was attempting to get at least one shaft spinning to enable her to egress the area.[45]

To prevent just such an escape, a strike launched from *Junyo* and *Zuikaku* arrived at 1500, unmolested by nonexistent air cover. Still having to brave the withering AA fire from *Hornet* and her escorts, Japanese pilots scored a torpedo hit on the carrier's starboard side and moments later a bomb penetrated the stern of the now-doomed carrier. *Hornet*'s skipper subsequently gave the order to abandon ship. Aware the Kondo's battleships could arrive in hours, Murray exiting into the coming darkness, leaving two destroyers behind to finish off the listing flattop before the enemy could have any opportunity to claim the floating hulk as a prize. Despite several spreads of torpedoes and gunfire, they only succeeded in setting the ship ablaze.[46]

At 2015, radarmen on one of the destroyers began to detect the oncoming Japanese Fleet to the west. Chased by Japanese floatplanes, the two tin cans departed to the east at flank, leaving the burning *Hornet* with 20 minutes of solitude before two Japanese destroyers arrived. Seeing that *Hornet* had a list of 45 degrees, it became apparent to Kondo that the carrier could not be salvaged. The two destroyers each shot two Long Lance torpedoes at the American carrier—kill shots causing *Hornet* to finally plunge into the deep abyss.[47]

CHAPTER 11

Engagement

With the loss of *Hornet* and *Enterprise* again needing repairs to mend battle damage, Task Force 64 now represented some of the higher playing cards in Halsey's hand. On paper, thanks to the outcome of the battle of Santa Cruz, it appeared the Japanese now held a much stronger deck as *Zuikaku* and *Junyo* remained in the region, supported by an impressive number of surface combatants and submarines. Yet the Japanese could not exploit this advantage. On the night Kondo's surface forces came upon the burning *Hornet*, he continued to be pestered by night-flying Catalinas. Rather than charge on, Kondo ceased pursuit and withdrew. Though two carriers remained, the number of available aircraft and crews to conduct follow-on operations had been greatly thinned out by an aggressive American combat air patrol and the increasingly effective antiaircraft gunfire from American combatants. With two flight decks already sent home to undergo repairs, the unscathed *Zuikaku* would also return to home waters to provide a training flight deck for aircrews undergoing flight training. Meanwhile Kondo's battleships could not be sustained at sea for an extended period given their fuel consumption. Thus, Halsey not only prepared to reshuffle his cards, but he also chose to continue playing them.[1]

As *Hornet* began her plunge to the bottom of the South Pacific, Lee's task force steamed south of Rennell Island to stay well out of the arc of Nagumo's carriers. For *Washington*, the early morning of October 27, a date that had been celebrated for two decades as "Navy Day," brought some good fortune as Lee's flagship *Washington* avoided the fate of sister

North Carolina when she avoided torpedoes fired by a lurking enemy submarine.

With reports placing some 20 Japanese submarines in the vicinity there was no surprise when at 0329, *Lansdowne*'s lookouts spotted a torpedo aimed at the battleship, sending crews to General Quarters throughout the formation. *Washington* successfully dodged the underwater missile and then faced another spread of Long Lances as dawn approached. Again alerted on time, the formation turned away from the oncoming spread to avoid the cataclysmic consequences. Since *Atlanta*'s station was beyond the battleship, *Atlanta*'s stern lookouts witnessed one of the torpedoes detonate between the two warships and another continue to follow in the wake, porpoising along the ocean's surface until it reached the end of its run time and sank about 800 yards from the light cruiser.[2]

After that eventful morning, the rest of the day was rather routine for Jenkins's crew as they prepared to embark a flag officer—making *Atlanta* for the first time—a flagship.

The next morning, *Atlanta* conducted a highline transfer with *San Francisco* to take aboard Rear Admiral Scott, his small staff, and Associated Press correspondent Tom Yarborough. One by one the newly arriving guests were each strapped into a basket-framed device and rode over under a line kept tense by heaving sailors. With Scott's arrival, Halsey established Task Group 64.2 under Scott's command as a new component of Task Force 64. Along with *Atlanta*, Scott assumed custody of the destroyers *Aaron Ward*, *Benham*, *Fletcher*, and *Lardner*.[3]

A graduate of the Naval Academy's Class of 1912, Scott first saw duty at sea in the battleship *Idaho*, and when the United States entered the first world war, he was the executive officer of the destroyer *Jacob Jones*, one of the destroyers to make the transatlantic crossing in May 1917 to Queenstown, Ireland, under Commander "We are Ready Now" Joseph K. Taussig. Operating in the Irish Sea, *Jacob Jones* recovered hundreds of victims from ships sunk by German U-boats until she herself fell victim to a torpedo fired by *U-53* killing 66 American sailors. Scott survived, in part, because the German U-boat skipper had the humanity to radio Queenstown on the fate of the first American destroyer lost to hostile action.[4] Rescued, Scott served as a junior naval aide to President

Atlanta, seen from *San Francisco* (CA-38), being refueled on October 16, 1942. Twelve days later the heavy cruiser transferred Rear Adm. Norman Scott to *Atlanta*. A photo of Scott likely during his command of USS *Pensacola* (CA 24) (Archives Branch, Naval History and Heritage Command, Washington, DC; NH 97807; 80-G-20823)

Wilson before rejoining the fleet to serve and succeed in assignments of increasing responsibility. At the outbreak of World War II, he held the rank of captain, in command of *Pensacola*, charged with escorting a small convoy en route to the Philippines. With the American base at Cavite shattered by Japanese air attack, the convoy changed course to Australia, and Scott was recalled to Washington. Promoted to flag rank in May, he received orders to return to the Southwest Pacific in time to witness the Marine landings on Guadalcanal. Being embarked on *Atlanta*'s sister ship *San Juan*, which protected the eastern approaches to Ironbottom Sound, spared Scott of the Savo Island debacle but positioned him close enough to learn from the numerous mistakes made that night. Placed in command of Task Force 64 by Vice Admiral Ghormley, Scott concentrated on training and nighttime maneuvers, anticipating actions aimed to disrupt the Tokyo Express. Despite the botched turning maneuver moments before the battle off Cape Esperance commenced, overall the preparation boded well for the Americans that night and Scott earned some accolades in the press back in the States.[5]

Despite the good ink in the media, Scott's transfer represented an unintended—and others would later say an unfortunate—demotion for the recent victor at Cape Esperance. Since Vice Admiral Halsey brought his own experienced staff to Nouméa, Ghormley's chief of staff, Rear Adm. Daniel Callaghan, found himself without work. A graduate of the USNA Class of 1911, and promoted weeks ahead of Scott, Callaghan had seniority on Scott. At the outbreak of war, Callaghan also had a cruiser command, witnessing the attack on Pearl Harbor from the bridge of *San Francisco*. Surviving the attack unscathed, Callaghan's ship deployed to support carrier strikes conducted on Japanese garrisons in early 1942. Promoted, Callaghan served in a pivotal position on Ghormley's staff. Upon relieving him, Halsey assigned Callaghan back to his former command to lead a cruiser surface action group.[6]

The arrival of Scott, who had experience with surface engagements, provided added reassurance to an *Atlanta* wardroom which understood that during her six months in the Pacific, the light cruiser had yet to fire on an enemy warship. Lieutenant Commander Mustin later reflected: "He knew exactly what his little force was and what he intended to do with it."[7]

Shortly after his arrival, Scott observed *Atlanta* steam alongside *Washington* to refuel, and afterwards the new task group commander ordered 17.5 knots and set course towards at Guadalcanal, assigning the four destroyers screening stations around his new flagship. Then Scott stationed *Lardner* on the near horizon to allow the remaining ships to conduct offset gunnery with their main batteries. Antiaircraft gun crews also honed their coordination by shooting rounds skyward. Scott observed some 90 minutes of gunnery drills before allowing *Lardner* to rejoin the formation which notched up its speed of advance to 19 knots.[8]

Having previous command of the heavy cruiser *San Francisco*, Rear Adm. Daniel J. Callaghan would use his former ship as a flagship despite the heavy cruiser not having the most capable radar. (Archives Branch, Naval History and Heritage Command, Washington, DC 80-G-20824)

Fortunately for Jenkins, the recently promoted flag officer was not put off by *Atlanta*'s spartan accommodations. Having earlier embarked in *San Juan*, Scott understood the light cruisers were not designed to serve as flagships. Also, Scott mitigated his need for spacious quarters below by spending most of his time seated on the port side bridge wing chair. The three rotating officers of the deck—Lieutenant Commanders Mustin, Smith, and Wulff—and other bridge watch standers found the junior admiral to be amiable and engaging.[9]

The next day the task group closed on Cactus and made course and speed adjustments to facilitate an early morning arrival off Lunga Point before dawn on October 30. On their way to their destination, Scott's flotilla steamed due north to pass the east side of the island that was spotted to the northwest at 2236. Before midnight the five-ship flotilla entered Indispensable Strait.[10]

Having completed their northward leg after midnight, the task group came left to a course of 295 degrees and slowed to 15 knots, forming a column led by *Aaron Ward*. Scott placed his flagship in the third or middle position as the formation approached Sealark Channel. Clearing that channel before 0400 placed Scott's small flotilla off Lunga Point in anticipation of the arrival of Marine spotters. From his position on the signal bridge, Ed Corboy watched American aircraft landing and taking off from Henderson Field: "Flashing shell bursts lighted the scene at intervals as the Marines and the Japs traded early morning punches."[11] Small craft from the Cactus Navy delivered Marines armed with maps marked up with suspected Japanese troop concentrations, artillery positions, and depots. On board *Atlanta*, Scott and Jenkins welcomed Maj. Charles M. Nees and his small detachment. A Philadelphian who signed on with the Marine Reserves as a student at Temple, Nees served as the operations officer for the 11th Marines. With Marine spotters embarked on all five warships, *Atlanta* took the lead to conduct a "hit and run" mission to pummel positions on Japanese-held portions of the island before Japanese air power from Rabaul could interdict.[12]

Firing commenced at 0629. Over the next two hours, *Atlanta* mount crews fired 4,093 5-inch rounds. The trailing destroyers with their single mount 5-inch guns aimed additional ordnance at the Japanese.[13]

In the early light, an Army Airacobra pursuit plane added an overhead set of eyes on the targeted enemy, diving at exposed targets and radioing adjustments back the task group. Working along the coast from the Matanikau River delta to Tassafaronga Point, *Atlanta*'s ammunition handling crews passed up a Whitman's Sampler of assorted rounds, emptying the magazines as Nickelson's gun crews rapidly fired the ordnance. Solid nose rounds with a delayed fuze shot penetrated bunkers and decimated foxholes. In contrast, antiaircraft rounds rained metal shards on exposed Japanese troops. Nickelson's gun crews even fired star shells, hoping to set off combustibles. Closing on the shoreline brought the enemy within range of the smaller AA guns. Graff observed, "It was carnage ... we blew up ammunition dumps, supply depots, and I'm sure a lot of people as well."[14]

At 0814, a suspected periscope sighting caused a ceasefire as the ships conducted evasive maneuvers to foil an anticipated torpedo attack. Believing the sighting to be false after eight minutes, shelling resumed. The ammunition handlers and gun crews probably appreciated the short break.[15]

As the guns fired, cartridge tanks—commonly called shell casings—that had contained the powder that exploded to launch the projectile out the gun barrel, were ejected out the backs of the engaged mounts and piled up. Graff recalled, "You couldn't walk on the main deck, there were so many shell casings." Upon declaration of the final ceasefire, Sears's deck hands used hoses to spray water to cool the now paintless gun barrels. Others grabbed the expended casings that would be stowed below to be eventually transferred off to be recycled.[16]

Ensign Corboy observed that Major Nees could not thank them enough as he departed, noting that tears welled up his eyes. The Marine visitors also left with cartons of cigarettes. "We had some kinds of cigarettes aboard the ship that nobody like ... so we gave them boatloads of these cigarettes" recalled Mustin.

At 0920, *Aaron Ward* led the now-departing column away from Lunga Point and sped at 28 knots to clear Skylark Channel before Japanese aviation could appear. By noon, having slowed to 25 knots, the column swung south to reverse passage of Guadalcanal's eastern shore en route for Espiritu Santo. At 1440 the next day, Scott's task group arrived and moored in Segond Channel. On *Atlanta*, the crew prepared to bring on fuel from the *Guadalupe* and take on the tiresome task of replenishing the light cruiser's near-empty magazines.[17]

Back on Guadalcanal, the Marines followed up the offshore barrage of enemy positions fronting them by pushing two battalions across the Matanikau River, meeting token resistance from the decimated Japanese defenders. Halsey, who delegated overall command of the naval forces near Guadalcanal to Rear Adm. Richmond K. Turner, committed to his subordinate to continue providing naval gunfire support. Turner complied, ordering Callaghan conduct an encore performance on the morning of November 4, forcing Japanese defenders to dodge 5-, 6-, and 8-inch shells fired from Callaghan's flagship *San Francisco*, the light cruiser *Helena*, and the destroyer *Sterett*.[18]

Though the Combined Fleet was no longer an immediate threat thanks to Kondo's decision to withdraw, the Tokyo Express continued to introduce new troops and supplies to reinforce the Japanese garrison. A day and a half following Callaghan's raking of Japanese positions, a light cruiser and 15 Japanese destroyers dropped off another regiment to fill the foxholes of comrades lost over the previous weeks. Yet, Admiral Yamamoto understood this continued piecemeal deployment of forces could only assure the stalemate. The Harvard-educated, twice former Japanese naval attaché to Washington, appreciated American industrial capacity and knew time was not his friend. The only path to achieve victory was to deliver overwhelming ground forces and deliver them soon.[19]

To do so he corralled an armada of merchant hulls and warships. On November 6, some 33 Japanese vessels were spotted less than 300 miles from Guadalcanal, parked near Shortland Island. Two days later, Halsey's headquarters received a report of a dozen transports gathering at the northern tip of Bougainville. While his staff in Nouméa sifted through the intelligence gathered by coast watchers, patrol aircraft, and from enemy radio transmissions, Halsey joined Vandegrift at Cactus to feast on Spam, dehydrated potatoes, and beans, and remain overnight to assess the situation ashore. A nighttime shelling by a passing Japanese destroyer gave the American naval commander a flavor for what his Marine compatriots had been enduring.[20]

The following morning on November 9, restocked with fuel, food, and ammunition, Task Group 64.2 departed from Espiritu Santo on an escort mission, having swapped out the destroyer *Benham* with *McCalla*. Up on *Atlanta*'s bridge, the task group commander worried that he was carrying the wrong bullets for an anticipated surface fight. With *Atlanta* having a primary air defense mission, most shells taken aboard at Espiritu Santo were fragmentation. Only 10 percent of the rounds received were hard-nosed Mark 32 variant. Writing to Halsey, Scott confided doubts on the effectiveness of AA rounds against enemy destroyers, not to mention cruisers or battleships. Given the Japanese advantages with their longer-range torpedoes, Scott argued the key to victory were cruisers with larger caliber guns capable of hitting the enemy before they could launch their fish and lamented that *Atlanta* was not ideally suited for such a surface fight.[21]

In the interim, to assure the Navy cargo ships *Betelgeuse*, *Libra*, and *Zeilin* reached Cactus unharmed, the main threats remained air attack and submarines. To foil the latter, Task Group 62.4 charted a course around the eastern and northern sides of San Cristobal island, the next island in the Solomons chain southeast of Guadalcanal. Callaghan's task group featuring the cruiser-types Scott pined for left Espiritu Santo a day later. Also en route to Cactus was another resupply convoy featuring four troop transports with *McCawley* serving as the flagship for the Rear Admiral Turner. The overall operational commander's transports were protected by the cruisers *Portland*, *Juneau*, and four additional destroyers.[22]

Having steamed the evasive counterclockwise route around San Cristobal island, Scott's four combatants and three cargo ships still were spotted by a Japanese seaplane during the daylight hours of November 10 as Guadalcanal still loomed in the distance. As with his earlier foray at the end of October, Scott's force conducted a nighttime passage along the northeastern coast of Guadalcanal, arriving off Lunga Point at 0530 on November 11. Small craft from the Cactus Navy rushed out to take on goods from the three cargo ships anticipating enemy aircraft would soon be in the vicinity. Sure enough, at 0938 nine Val dive-bombers escorted by Zero fighters made a run at the small flotilla. Scott, having been tipped off about the incoming raiders from shore-based radar, halted the off-loading operations and got his ships underway on a due north heading, forming a column led by *Atlanta* followed by *Betelgeuse*, *Libra*, and *Zeilin*, and four destroyers, flanking the column with two positioned on each side. Looking up, from their General Quarters stations, *Atlanta*'s smaller caliber gun crews witnessed the enemy dive-bombers flying in a line abreast and then Val on the right-hand end of the line pushed over and began his descent, with the other eight aircraft following his lead.

Since the cargo ships were the objective, *Atlanta*'s aft batteries, directed by the Assistant Gunnery Officer Lieutenant Commander Mustin, were best positioned to take on the oncoming attackers. The configuration of the light cruiser's gun mounts meant that five of the eight twin 5-inch 38 mounts could be brought to bear on the incoming Vals. The gauntlet of fragmentation rounds thrown up by *Atlanta*'s 10 5-inch gun bores, added to the rounds fired from the destroyers and from guns placed on the three cargo ships proved to be too much of an obstacle for the Japanese to overcome. That said, their rapid descent with mountains in vicinity challenged *Atlanta*'s fire-control radar which provided less than optimal readings. Instead, Mustin relied on the optical range finder to determine the elevations of the oncoming aircraft to allow the fuzes of the fragmentation shells to be mechanically set.

Mustin estimated they were able to focus his aft batteries on six of the nine dive-bombers, as the aft gun director could manually slew all of the guns onto a diving enemy plane, follow the plane down, and then slew the guns up to target another attacker. Because of that systematic approach of following down one aircraft and slewing up to the next, Mustin could not make claims on the number of aircraft *Atlanta* may have brought down.[23]

Overall, the defending AA fire knocked about half of the attackers from the sky. Mustin later argued misses also hindered the enemy pilots' focus and ability to put bombs on target as demonstrated in this attack where no direct hits were scored. However, three misses near *Zeilin* caused a hold to flood. As the remaining Vals attempted to clear the waters off Lunga Point, Marine Wildcats caught up, shooting each one down as well as some of the escorting Zeros.[24]

The off-loading continued until midday when a larger Japanese air strike consisting of three sections of Bettys approached. Not knowing the enemy objective, Scott again disrupted the transfer of goods to weigh anchor to repeat the earlier morning defensive maneuver. As the three cargo ships and five escorts again steamed northward, *Atlanta*'s Executive Officer Commander Emory and the backup shiphandling team posted at Battle II, located just forward of mount six, looked up at the incoming formation of three Vs of twin-engine bombers. Altitude proved an effective shield as defending Wildcats had trouble reaching 28,000 feet. Meanwhile, the upper limit of *Atlanta*'s twin mounts reportedly was 25,000 feet. Once it became apparent that the Japanese bombers were not targeting Scott's flotilla, Captain Jenkins turned *Atlanta* to port to unmask all but one of his eight mounts.

Down in Plot, Lieutenant Shaw kept reporting readings from the Mark 1 computer as placing the enemy aircraft beyond range. Lieutenant Commander Mustin asked for the maximum setting for the fuzes. Shaw responded: "45 seconds." So instead of asking for altitude calculations, Mustin asked for fuze calculations. As the bombers flew on a course parallel to *Atlanta*'s port beam, Shaw kept reporting the number of seconds of flight time it would take before a shell could burst within the enemy air formation. Mustin patiently waited as the second count came down. When Shaw called up 45 seconds, Mustin ordered "commence firing."[25]

Once Mustin's aft mounts fired, Bill Nicholson's forward mounts joined in. Unfortunately, as the Japanese bombers made their closest approach at midday, it meant they were also passing in front of the sun, making it near impossible to spot bursts and make corrections. During the minute that the Bettys passed within *Atlanta*'s maximum range, the light cruiser's gun crews were able to fire approximately two dozen rounds per barrel. Mustin peered up into the sunlight to assess the impact of his ship's ordnance and counted two Bettys falling away from their formations.

In his volume *The Struggle for Guadalcanal*, World War II naval chronicler Samuel Eliot Morison dismissed the attack as another attempt to disable

Henderson Field. However, the Japanese objective had been the area where the Cactus Navy had just landed the newly arrived cargo. Yet, in choosing to bomb at high altitude, the Japanese had traded off accuracy for safety. Few bombs hit their intended targets. One consolation for the Japanese was the attack succeeded in limiting American air operations for a short period as the metal shards from over 300 bursting shells rained down on Henderson Field.[26]

Following the midday raid, the three cargo ships again dropped their hooks, and small craft again ventured out to receive the welcomed supplies. *Zeilin*, with a flooded hold, became a liability that no longer needed to stay in the vicinity. Scott ordered the cargo ship to slip away under cover of darkness and assigned *Lardner* as her escort. As *Zeilin* and *Lardner* faded away to the southeast, Scott took his four remaining warships into the Indispensable Strait north of Lunga Point. There, at about 2200, they rendezvoused with Callaghan's force of cruisers and destroyers. Assuming the role as the overall tactical commander, Callaghan left the three destroyers to guard the two remaining cargo ships and had *Atlanta* join his formation to steam conduct two thorough sweeps of the approaches around Savo Island with the objective of disrupting any Tokyo Express arrivals.[27]

For the crew of *Atlanta*, having spent the previous evening at their battle stations and then having fended off two air attacks earlier that day, it may have been fortunate, given their sleep deprivation, that they did not come across any enemy units. Yet, in spite of the long hours at their GQ stations, Mustin noted the crew did not seem stressed and the men maintained good spirits. A steady delivery of food and water from the galley satisfied appetites, and sailors took turns to slip away to make needed head calls. On the bridge, Jenkins stayed calm and maintained just the right demeanor. Mustin observed: "He was a thoroughly practical leader and commander and wonderful one."[28]

Returning from their westward sojourn, Callaghan and Scott then met the small convoy centered around Rear Admiral Turner's four transports that arrived just before dawn. As the sun rose, Japanese observers on the overlooking hills could now see six noncombatants off the American-controlled shoreline. While *Betelgeuse* and *Libra* remained moored a mile east of Lunga Point, Turner's four troopships parked off Kukum Beach. A combined flotilla of small boats from the four transports and the Cactus Navy quickly began to ferry replacement Marines, additional Army battalions, and the stores needed to sustain them ashore. Reacting to the arrival of the unwanted guests, at 0718, the Japanese artillery piece dubbed "Pistol Pete" took aim at *Betelgeuse* and *Libra*. With splashes landing in the vicinity of the two ships, American gunners ashore as well as gunners on the destroyers *Barton, Shaw*, and the cruiser *Helena*, guessed where the gun was and fired rounds into the jungle canopy. On *Atlanta*, Mustin saw the offending gun through a clearing and sought permission to engage. He failed to get a response. Mustin later suspected that the situation with three admirals in the vicinity had an impact on command and control. The admiral embarked in *Atlanta*, Scott, was the junior of the three.[29]

Meanwhile trouble brewed to the northwest. To lead the forthcoming Japanese thrust to seize Guadalcanal, Admiral Yamamoto ordered a surface group centered around two battleships to steam down the Slot to pummel Henderson Field with heavy-caliber shells. Unlike an earlier bombardment where the Japanese battleships had evaded Allied detection, Turner received patrol plane sightings of either two Japanese battleships or heavy cruisers, a cruiser, and six destroyers en route down the Slot having an estimated arrival time later that evening. Rather than circling his combatants to defend his transports, the overall Tactical Commander Turner planned to place his cruisers and destroyers under Callaghan to head off the inbound Japanese battlewagons to impede their mission of pulverizing Henderson Field. Scott—the veteran of Cape Esperance—and his small staff, would have no command role for this forthcoming engagement.[30]

The Japanese, understanding that American surface forces sighted in the vicinity of Lunga Point could disrupt the bombardment mission, massed a small armada of bombers armed with torpedoes and launched them from Rabaul with an arrival time slated for just after 1400. Forewarned of the pending attack, Turner repeated the drill the Scott had perfected on the previous day, halting to off-loading operation to get his six transport and cargo vessels steaming in two columns of three in a northwesterly direction, having the combatant ships forming a protective barrier. *Atlanta*'s placement on the port side of the formation proved unfortunate in that the formation of 21 Bettys, escorted by Zeros, swooped down from the direction of Florida Island to approach Turner's formation from the starboard beam. Scott, Jenkins, Emory, Nickelson, Mustin, and the other topside personnel struggled to look around the two columns of noncombatants and the screening combatants stationed further starboard including *Atlanta*'s twin *Juneau*, as the attacking twin-engine enemy bombers split into two groups to ensnare the defending Americans with a crossfire of incoming torpedoes. One group veered right to circle in from the northeast and the other turned left to wheel in from the southeast. Each group approached in a line abreast that tightened as they closed on their release point. On *Atlanta*, Nickelson and Mustin grimaced as the other ships in the formation obstructed clear shots at the low-flying Bettys. Searching the skies for other targets, *Atlanta*'s lookouts spotted aircraft diving down following the Bettys. Thinking they could be enemy dive-bombers, Mustin told his aft gun crews to stand by to open fire.[31]

Upon closer inspection, Mustin realized they were witnessing Marine Wildcats and Army Airacobras coming to the defense of Turner's ships. Sixteen aircraft, split evenly between the two services, went after the Bettys coming from the southeast. Leaving those Bettys to the hands of the Cactus Air Force, Turner's formation on its northwesterly heading presented an inviting broadside to the line of Bettys closing from the northeast. Turner reacted, ordering a ninety degree turn to port, suddenly giving the northeast attackers an ass-end perspective of their intended targets. For the disappointed pilots of this wing of Bettys, the consolation was that Turner's ships now presented a broadside for their compatriots flying in from the southeast.

With the formation turned to the southwest, *Atlanta* now steamed at the head and those at their GQ stations on the port side witnessed the short work of the Marine and Army fliers on the incoming Bettys. With American fighters mixing it up with the incoming bombers that were now flying evasively for survival, *Atlanta*'s gunners continued to withhold fire. From his battle station in the forward gun director Lt. Pat McEntee admired the determination of one Marine pilot who, having expended his ammunition, flew over, lowered his landing gear, and pushed down on the fuselage of a Betty, forcing the plane into the waves below.

As McEntee watched the Cactus Air Force break up the attack from the southeast, Mustin—from his station just abaft of *Atlanta*'s aft gun director—experienced a panoramic view of the flotilla with the line of Japanese bombers approaching in the distance. He observed: "In they came and a fairly awesome sight it was." Coming to within 5,000 yards of the retreating American formation, the determined Japanese attackers witnessed salvos coming from stern main batteries of the cruisers *Helena* and *San Francisco*, that smacked into the sea ahead of them, causing huge geysers which they had to evade. Then came in the 5-inch and smaller caliber fire thrown out by *Juneau* and the other ships on the back end

Smoke rises from two enemy planes shot down during a Japanese air attack on U.S. ships off Guadalcanal, November 12, 1942. Photographed from USS *President Adams* (AP-38). (Archives Branch, Naval History and Heritage Command, Washington, DC; 80-G-32367)

of Turner's formation. Some of the Japanese pilots instinctively pulled back on their sticks to evade the incoming barrage but in doing so only gave the American gunners a clearer shot at their undersides.[32]

One by one Bettys plunged into the Sound. Those that managed to fly over Turner's formation after dropping their fish had to face *Atlanta*'s after batteries. Mustin recalled: "We had a chance to fire at about two, perhaps more, but at least two that we fired on we saw go down in the water between us and Guadalcanal."[33]

Though no enemy torpedoes reached their intended targets, the Americans did not escape unscathed. The pilot of one crippled Japanese plane, after targeting his torpedo against *San Francisco*, piloted his aircraft at the heavy cruiser. Though the torpedo missed, the pilot veered his plane into the aft structure housing the control station for the rear triple 8-inch gun turret. In addition to losing the after control station, the suicidal attack damaged the aft antiaircraft director and radar, and destroyed three 20mm mounts, killing their gun crews. Aviation gas splattered and ignited, causing additional casualties including severe burns on the XO's legs. Rear Admiral Callaghan from his position on the forward flag bridge, watched as damage control teams reacted to the blow.

In addition, five sailors on *Buchanan* were killed and seven more wounded when the destroyer's aft stack was struck by a not-so-friendly 5-inch round. With the departure of the raid, the formation returned to its previous location off Lunga Point to recommence the off-load of men and material. As *Atlanta* steamed back, the crew looked over to witness floating fuselages and chunks of fallen aircraft.[34]

CHAPTER 12

Friday the 13th

On the afternoon of November 12, the rush continued to empty the holds of the six transports and cargo ships as they had been given an eviction notice to vacant the premises. Before dusk, all of the American ships off Lunga Point—Turner's Task Force 67, including *Hovey* and *Southard*; the two World War I-vintage former destroyers converted to high-speed minesweepers—weighed anchor and departed eastward through Sealark Channel. The timing of the departure of the whole formation was a deceptive move for the benefit of Japanese observers on the overlooking hills who would hopefully report up to Rabaul that the waters off Guadalcanal had been cleared of enemy warships. Arriving at the Indispensable Strait, the formation split up with Rear Admiral Turner embarked in *McCawley*, the five other noncombatants, *Hovey* and *Southard*, and destroyers *Buchanan, McCalla,* and *Shaw* continuing to Espiritu Santo where they would arrive on November 15. Rear Admiral Callaghan, who remained with the damaged *San Francisco*, led Task Group 67.4, the support group composed of the remaining combatants, back through the more southerly Lengo Channel.[1]

To navigate through the channel and form a disposition to challenge the oncoming Japanese, Callaghan formed a 13-ship column with his eight destroyers evenly bracketing his five cruisers. *Atlanta* led the five cruisers, followed by *San Francisco*, *Portland*, *Helena*, and sister *Juneau* in a single-file column. Destroyers *Cushing*, *Laffey*, *Sterett*, and *O'Bannon* led the formation while destroyers *Aaron Ward*, *Barton*, *Monssen*, and *Fletcher* followed in the cruisers' wake. Approximately 500 yards separated each of the destroyers while 700-800 yards were maintained as the distance between the cruisers. When adding the separation distances and ship lengths together, Callaghan's column stretched some four miles. For those contemplative watch standers steaming through the night, the supposition of going into battle with 13 ships as a component of Task Force 67 (which also totaled up to 13) as the calendar turned to Friday the 13th seemed foreboding. That dark night's partial overcast sky with flashes of lightning arcing over Guadalcanal and Florida Island also added to the tension.[2]

Callaghan's column steamed westward at 18 knots, passing north of the Lunga Point beachheads. As the island coastline curved in a northwesterly direction, adjustments were made to the column's heading to keep the formation parallel to the coast.[3] Meanwhile the oncoming Japanese force had consolidated earlier that afternoon when Vice Adm. Hiroaki Abe's bombardment group, centered around the battleships *Hiei* and *Kirishima*, rendezvoused with Rear Adm. Tamotsu Takama's destroyer force to proceed on the final leg to Guadalcanal. En route, Abe nearly called off the mission as his force encountered several drenching squalls that hindered his ability to maneuver, let alone place 14-inch shells on target. Reassured by Japanese transmissions from Guadalcanal of no precipitation in the vicinity, Abe radioed Takama and his destroyers to surge forward to scout the waters off Lunga Point. However, steaming earlier through the near-zero visibility conditions of the squalls, Takama's van had managed to fall back to Abe's rear. With no signals to the contrary, Abe assumed Takama's unseen destroyers were still ahead, clearing the waters in front of the bombardment group. Not hearing of any sightings ahead, at 0130 Abe ordered his gun crews to prepare for shore bombardment. From the magazines below, the ammunition handlers broke out incendiary projectiles and high-explosive shells that were hoisted up into the turrets of his battleships.[4]

On the *Atlanta*, Electrician's Mate 3rd Class Bill McKinney had just come off his 2000–2400 watch in the after engine room. With no anticipated immediate threat, Captain Jenkins kept the ship at Condition 3 to allow crewmembers to get some necessary shut-eye before the anticipated clash. The courtesy did little for McKinney. After taking a

The Japanese battleship *Hiei* served as Rear Admiral Tanaka's flagship. (Archives Branch, Naval History and Heritage Command, Washington, DC; NH 89173)

shower, he entered his red-lit berthing compartment located four decks down ahead of the forward fire room just over the magazine for the 5-inch 38 mount three. Just as he climbed into his bunk at about 0100, General Quarters sounded. As McKinney's shipmates climbed out of their berths to don shoes and outer garments, and rushed out to their GQ stations, McKinney did not need to go far as his berthing space which also doubled as a damage control substation for Repair II. In addition to numerous bunks, the interior space contained a locker loaded with sledges, chocks, hoses, and other damage control equipment including two sets of Rescue Breathing Apparatus (RBAs) for use in entering smoke-filled spaces. McKinney, joined by Seaman 2/c Daniel Curtin, put on his sound-powered headset with chest-mounted speaker and reported to Repair II "Manned and Ready."[5]

As McKinney and Curtin waited at their interior battle stations, Vice Admiral Abe surged forward not fully understanding the whereabouts of all of his ships in his disorganized formation. In contrast, many of McKinney and Curtin's compatriots observing radar scopes decks above knew exactly where Abe's ships were. *Helena*'s superior SG radar gave its operators an exceptionally clear picture of the enemy force. Even on *Atlanta*, Scott and Jenkins were kept informed about the surface contacts

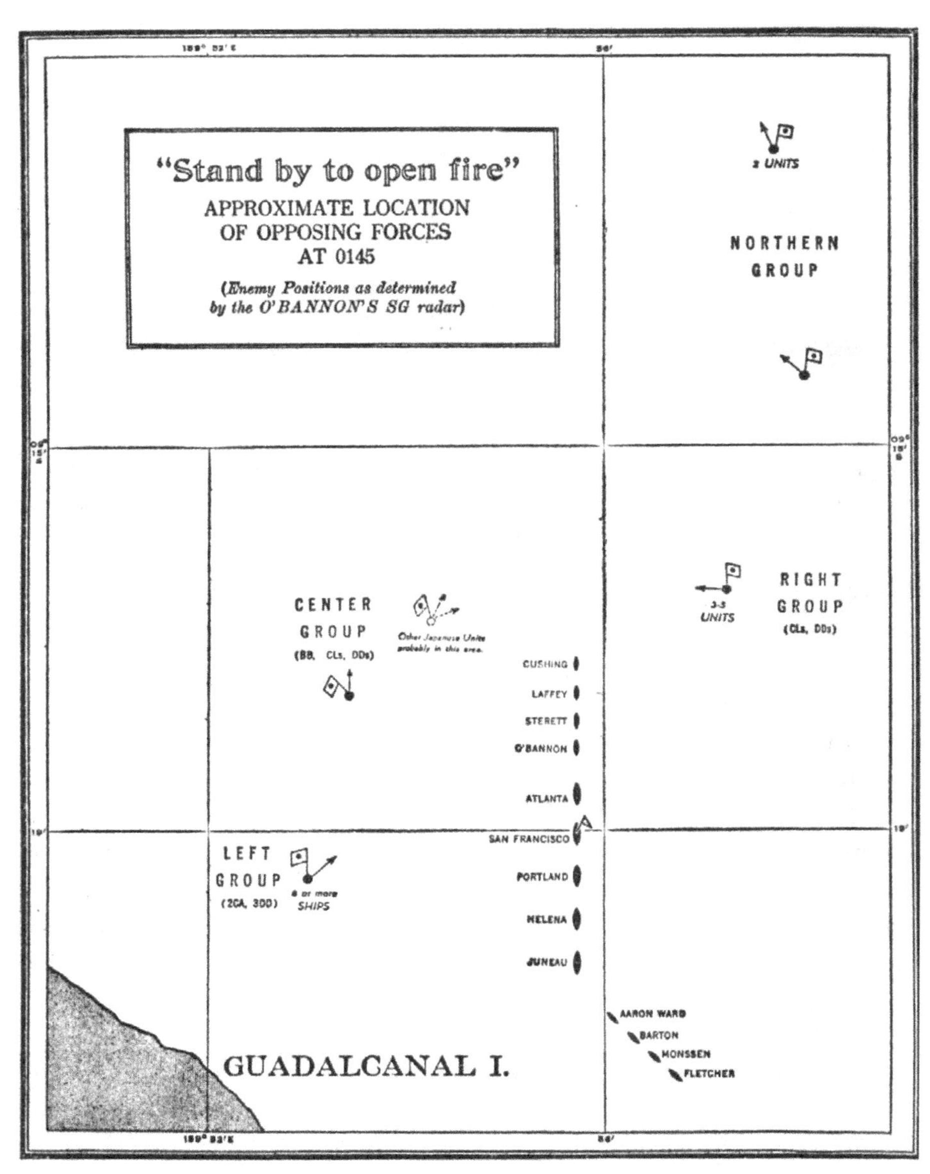

Chart from the Office of Naval Intelligence *Combat Narrative of the Battle of Guadalcanal*, published in 1944. (*Solomon Islands Campaign: VI—Battle of Guadalcanal, 11–15 November 1942*, Office of Naval Intelligence, United States Navy: Publications Branch, 1944, p. 18)

11 miles distant to the northwest which had begun appearing on the light cruiser's less capable SC set. Lieutenant Commander Mustin recalled. "These were big targets and they showed up on our radar." *Helena* apprised Rear Admiral Callaghan through radioing radar contact reports to his flagship. As the reports came across TBS, Callaghan ordered the column to pivot to a direct north heading.[6]

If the intent of the pivot was to cross ahead of the path of the oncoming Japanese, the order came too late. With *Helena* tracking the Japanese advancing at 23 knots, the two formations closing rate of 40 knots meant Callaghan's lead ship, some two miles ahead, was steering the formation directly into the path of the Abe's warships. The limited ability of American lookouts to pick out the approaching enemy warships, thanks to the presence of nearby squalls, only further squandered any tactical advantages gained through the possession of radar.

As *Atlanta* turned to starboard to follow in *O'Bannon*'s wake, her forward and aft gun directors focused on targets off the port bow. From his perch aft of the rear gun director, Assistant Gun Boss Mustin called down to Plot obtain readings from the two Mark 1 computers, each one linked to a director. Looking at the readings, Lieutenant Shaw confirmed the fore and aft directors were tracking two separate targets.

Because Callaghan had promulgated the message "Stand by to commence firing," the 13 ship captains were obliged to withhold fire on the oncoming unsuspecting enemy despite the many resolved fire-control solutions up and down the line. Throughout *Atlanta*, sailors waited. Back in turret eight, Clifford Dunaway stood ready as a loader. He later admitted he was scared to the point he was shaking but took some solace when his misinformed mount captain passed along a report that there were only three enemy ships up ahead. With the American column having 13 ships, this should be a one-sided affair, he thought. But he was also one of those superstitious crewmembers who winced at the notion that they were going into battle with 13 ships on Friday the 13th, and that mount eight was crewed by 13 sailors. Further forward and deeper down, Bill McKinney waited to respond to the impact of enemy ordnance. As he sat, he felt the vibrations of *Atlanta*'s engineering plant shudder through the bulkheads. Surging towards combat, the Engineering Officer,

Lieutenant Commander Loeser, turned off the ventilation system as a precaution against spreading smoke throughout the ship. As the roar of air circulating through the shop ceased, McKinney's ears trained on the sounds of the ammunition hoists that were hauling up 5-inch common round shells from the magazine a deck below to a twin mount above.[7]

For Nickelson and Mustin, the Mark I computer-generated fire-control solutions no longer mattered. Once the range closed on a target the size of a ship, all the gun boss and his assistant needed to accomplish was aim at the target and, once the director locked on it, start shooting. Since the gun mounts were placed in automatic control to follow the motions of the directors, the gun crews within the steel boxes struggled to keep standing let alone feed shells into the guns during rapid slewing.[8]

At 0142, the lead ship, *Cushing* came to port after spotting enemy destroyers crossing ahead to unmask both her torpedo tubes and aft gun mounts to starboard. Having received a request from *Cushing*'s commander to open fire, Callaghan replied with instructions to return to the formation heading. However, the destroyer's skipper turned even further to port to avoid colliding into the two crossing Japanese destroyers. *Cushing*'s turn had a cascading effect. A collision threat also forced *Laffey*, *Sterett*, and then *O'Bannon* to order even sharper hard-left rudders to avoid running into the ship ahead. By now, lookouts on the two Japanese destroyers could not but help to notice the churning water and the ships that were causing it off their starboard beams. Having dodged the Japanese destroyers, the van destroyers each turned starboard to cut across the bow of the oncoming *Hiei*.[9]

At this time *Atlanta*'s aft director tracked a target to starboard beyond the two passing Japanese destroyers. Mustin peered into his binoculars and spotted the three-stack silhouette of a two-decade old Japanese light cruiser that he would later learn was *Nagara*. Armed with Long Lance torpedoes and seven 140cm (5.51-inch) guns, *Nagara* offered an inviting target. As Mustin studied the enemy hull, he heard Shaw's voice in his sound-powered headset reporting up from main plot that the light cruiser was a mere 6,000 yards away.

As Mustin kept his binoculars fixed on *Nagara*, up on the bridge Captain Jenkins witnessed *O'Bannon*'s turn to port that caused her to fall back towards the oncoming column. To avoid collision, Jenkins ordered a hard-left rudder. Throughout the ship, especially up on the signal bridge and gun directors, crewmembers hung on as the light cruiser heeled hard to starboard. Callaghan, embarked in the trailing *San Francisco*, seeing his lead cruiser veer to the left, radioed Jenkins, "What are you doing, Sam?" *Atlanta*'s captain retorted, "Avoiding our own destroyers."[10]

Callaghan answered, "Come back to your course as soon as you can. You are throwing whole column into disorder." *Atlanta*'s turn to port placed the light cruiser on a reciprocal heading of the oncoming Japanese. Mustin kept his mounts fixed on *Nagara*, awaiting an order to open fire.

Suddenly a powerful searchlight lit up *Atlanta*. The powerful beam from abaft the light cruiser's port beam, likely emanated from the Japanese destroyer *Akatsuki*, ruined the night-adjusted vision of those assigned to GQ stations along the port side of the ship. Mustin remembered, "This searchlight seemed so powerful and high up in the air as seen from my position in the after air defense station that I thought it was a big ship with a huge searchlight located way up in the superstructure, as cruiser of battleship."[11]

Startled, Captain Jenkins ordered his signalmen counter illuminate with *Atlanta*'s searchlight. However, his gunnery and assistant gunnery officers instinctively communicated an alternate response. Bill Nickelson yelled "Fuck that, Open Fire!" to his forward gun mounts. Back aft Mustin ordered "Action Port." The aft gun director quickly slewed around from starboard to port in the direction of the offending light. Mustin called

This image of *Akatsuki* was taken months earlier in the Aleutians. (Archives Branch, Naval History and Heritage Command, Washington, DC; NH 73049)

down to Shaw in the plotting room to report an estimated range of 2,000 yards. Once the gun director officer reported on target, Mustin ordered "commence firing."[12]

Remarkably, the Japanese destroyer that decided to illuminate *Atlanta* had yet to open fire on the ambushed light cruiser. That indecision proved costly as *Atlanta*'s aft batteries were quickly followed by the forward mounts in quickly dispensing 5-inch shells towards the target. Back in mount eight, Seaman Dunaway braced as the turret had whipped around and when the round in the breech fired and was ejected, his nerves suddenly calmed, and he began quick feeding the 52-pound rounds into his gun's breech loader as fast as the magazine below could feed them up.

From his perch, Mustin saw some shells splash into the water short of the target while others passed overhead. Both the forward and aft gun director officers adjusted aim. Tracer rounds could now be seen penetrating the target. By the time *Atlanta* closed to 600 yards of the enemy destroyer, it was becoming target practice. Mustin observed: "Shooting into a destroyer size hull at 600 yards you just don't miss … you just don't miss." By this time the Japanese captain either had ordered his signalmen to extinguish his searchlight or had it turned off for him by an incoming round from *Atlanta* or other American ships that followed *Atlanta*'s initiative to engage the enemy. Rear Admiral Scott, from his port side perch would have observed 6-inch rounds from *Helena*, followed by 8-inch shells fired by *San Francisco* on the offending destroyer.[13]

The task group commander immediately recognized the folly of this piling on action as the mass fire on one ship exposed the engaged American combatants to counterbattery fire and torpedo attack. Hoping to resolve the problem, Callaghan ordered "Odd ships commence fire to starboard, even to port," a command that caused confusion up and down a column which had already begun to disintegrate thanks to the collision avoidance maneuvers of the lead ships.

Atlanta's rapid-firing 5-inch twin mounts shooting in a direction opposite of the main body of Abe's force, combined with star shells fired from the two battleships, silhouetted the light cruiser. One of the Japanese advantages of night-fighting was they still used gunpowder that produced heavy smoke, a disadvantage in daylight but at night the brown clouds covered and masked muzzle flashes. In contrast, American smokeless powder exposed explosive balls of flame for all to see. Two Japanese destroyers that had been trailing *Akatsuki* emerged from the cloak of a rain squall to view *Atlanta* as if it were daytime. *Inazuma* launched a spread of six torpedoes.[14]

Meanwhile, Japanese light caliber shells, likely from *Akatsuki*, ripped into *Atlanta*'s torpedo control killing Lt. (jg) Henry Jenks as he prepared for launching. Then a torpedo penetrated *Atlanta*'s forward engine room, close to the bulkhead, separating that space with the forward fire room.[15]

The blow startled Seaman 1/c Franklyn Reed who stood just outside the pilothouse as Rear Admiral Scott's personal messenger: "What the hell was that?" Admiral Scott, who had experienced the loss of *Jacob Jones* to a German U-boat torpedo during the previous war, responded: "That's a torpedo."[16]

The explosive blast below the waterline shot up a huge geyser of oil-infused water and quickly snuffed the lives of those stationed in the forward engine room including the Chief Engineer Lieutenant Commander Loeser, Chief Machinist's Mate Henry Wolfe, the throttleman Machinist's Mate 1/c Henry Merrill, the feed pump monitor Machinist's Mate 1/c James "Pop" Crawford, the generator and switchboard monitor Electrician's Mate 2/c Thomas Tuck, and phone talker Fireman 1/c Steven Kiraly.[17]

With the design and construction of the super dreadnought *Nevada* prior to World War I, American ship designers took an all-or-nothing approach regarding the installation of armor to protect vital versus nonessential spaces. In *Atlanta*'s case, this design philosophy contributed to her eventual demise as armor overhead and along the sides of the hull channeled the explosive energy of the blast against the weaker fore and aft bulkheads that encapsulated the space. In addition to destroying the forward engine room, the blast and subsequent flooding knocked the forward boiler room out of action, and still had enough energy to rupture the bulkhead between the forward fireroom and the plotting room.[18] Though armored with three inches of steel, the ceiling directly over the engine room also buckled with fatal results. The compartment above, the forward mess hall, served as Repair V during General Quarters. The force of the blast plastered the hall's massive steel serving table into the overhead and wiped out the five-man repair party that had been led by Lieutenant (jg) Pierce and Chief Machinist's Mate Anthony Arrigoni.[19]

For the five men stationed in the forward fireroom, all survived the shock of the blast in the adjoining compartment to scamper up the ladder as the space began to flood. Chief Watertender Thomas Maloy, made it up to the second deck escape hatch and scampered back to the forward engine room to attempt to save those trapped, a noble effort that apparently cost him his life. Fireman 1/c Joseph Harvey was the last out with water up to his armpits as he reached the escape ladder. Tumbling out of the hatch, Harvey followed three of his shipmates forward through the library into Repair II where Lieutenant (jg) Nicholson led a repair team preparing to spring into action along with Lucky, the ship's mascot. Timing proved unfortunate for the forward fireroom survivors as a shell entered the space killing all (including Lucky) except for Harvey who managed to emerge unscathed from under the bodies of his shipmates. Shortly afterwards, Electrician's Mate Oscar Ekberg entered the space from below where he and two others had been assigned to monitor the interior communications room which now was useless without power. The gory image he saw would be forever etched in his memory.[20]

In the plotting room, the torpedo detonation threw Lieutenant Shaw against a bulkhead, smashing his right hand. Shortly thereafter, water began to pour into the compartment and Shaw told Lieutenant (jg) Wilson and another man assigned to the space to escape topside while he and the remaining two sailors tried to stem the incoming flow. Apprising Lieutenant Commander Nickelson of the situation by his sound-powered phone, Shaw was told to "stick a pillow in it." The line went dead. Shaw needed more than pillows as salt water quickly rose. Praying he would not drown, he told the remaining two men to head topside through an escape trunk. With his smashed hand, he clambered up the vertical tunnel leading to an egress by the quarterdeck. En route, another blast threw him into another bulkhead knocking him unconscious.[21]

Since the torpedo detonation also ruptured the bulkhead between the forward engine room and after fire room, water began to also pour into that vital space. The watertenders had little choice but to cut off the black oil feeding the boilers and, as a final step as they escaped from the space, lifted the safety valves that enabled built-up steam to escape up through the aft stack. Fireman Kenyth Brown was the first up the ladder and tried to open the hatch with no luck. Fortunately, Watertender 2/c Ignatius Burakowski who had been monitoring the evaporators on the second deck, used a battle lantern as a hammer to pound on the dog handles to get the hatch open. In addition to Brown, Chief Watertender John Mercer, Fireman 3/c Wayne Langton, and two other terrified shipmates emerged from the nearly flooded compartment. Brown later recalled the "tremendous feeling of relief" he felt when the hatch opened and Burakowski helped him out.[22]

The loss of both boiler rooms cost *Atlanta* her source for propulsive power, and the stricken light cruiser slowly drifted to a halt. As steam drove the ship's electric generators, and electricity powered the 5-inch 38 twin mounts, the fire-control radars, etc., *Atlanta* became defenseless as additional incoming rounds continued to punish the ship.

The release of steam through the aft stack crippled the ability of Commander Emory and others assigned to Battle II to communicate. Nearby, Mustin seeing two clusters of chemical smoke tanks located on the stern thought: "Let's get those smoke tanks to work and cover us in a cloud so that every Tom, Dick, and Harry won't see those fires and take a few shots at us."[23]

Mustin proceeded to open the valves located at his after control station, sending compressed air back to the tanks to force the chemical compound into the atmosphere to make thick smoke. To his dismay, nothing happened. Suspecting enemy shellfire had cut into the lines Mustin grabbed, Lt. (jg) Al Newhall, the aft director officer, who had just climbed down from his perch. With steam roaring out the aft stack thanks to the lifted safeties, Mustin pulled Newhall to the base of the aft director, slammed shut the hatch behind him, and after forming his hands into a cup around Newhall's ear, yelled at the junior lieutenant to grab a pistol to run back aft to shoot into the tanks. To Mustin's frustration, "He never did get the message."[24]

While the torpedo hit had not caused fires, the shellfire hits did. One shell scored a centerline direct hit on the forward gun director and blew inward to the empty void below the director that had served as a lookout station, killing all who had GQ stations in that cramped space including Lieutenant (jg) Rhodes. The blast set fire to a mass of wiring that ran down conduits leading to the plotting room. The Director Officer Lt. Pat McEntee stationed above, had no egress. With *Atlanta* taking on a port list, he had only one option … he plunged overboard into Ironbottom Sound.

Additional enemy gunfire raked the 1.1-inch and 20mm batteries forward on the port side killing Ensign Straub, Chief Gunner's Mate John Konecny, Seaman 1/c John Johnson, Seaman 1/c Enrico Taormina, Seaman 1/c Clark Davis, Seaman 2/c John J. McKee, Seaman 2/c Lloyd DeHerrera, Seaman 2/c Donald Labonte, Seaman 2/c Kachazooney Chakarian, Seaman 2/c Wesley Belaire, Seaman 2/c Charles Cruse, and Seaman 2/c Keith Crawshaw. As a mount captain of a 1.1-inch battery on the unengaged starboard side, Boatswain's Mate 1/c Spadone had his crew take shelter against their splinter shield as metal fragments whizzed through the air "like dried beans in an aluminum pot." With half of his gun crew wounded by shrapnel including himself, Spadone performed immediate first aid.[25]

At their GQ station deep below, up forward, McKinney and Curtin could hear *Atlanta*'s batteries unleash and then they were knocked over when the port side torpedo hit jolted the ship. As electrical power dropped offline, battery-powered battle lanterns came on in interior spaces throughout the ship. Now what? McKinney was getting no instructions and training dictated they not open any hatches that could endanger the ship's watertight integrity. McKinney could hear tearing noises as well as coughing and choking. Clanging a spanner wrench on the hatch leading to the next compartment forward, McKinney got a clang back from a shipmate who informed him that the forward corner overhead of that compartment had been blown open and they could see flames, and blood was dripping down. They were getting out. Rather than opening that hatch, McKinney decided to open an escape scuttle that had been placed in a clamped-down overhead hatch. Seeing thick yellow smoke, McKinney closed the scuttle hatch back down and reached for the RBAs. Having never used one, McKinney quickly read the instruction booklet under the light of a battle lantern; put the harness holding the breathing contraption on; grabbed a canister, tearing off its seal, and shoved the canister into a housing below the mask; pinched on rubber hoses leading to his mask; and inhaled and exhaled into the rubber bags positioned on each side of the canister housing. After some struggling, Curtin also got his RBA to function. Climbing into the space above, McKinney and Curtin made their way through the smoke-filled compartment, stumbling over the body of a fallen shipmate and then proceeded up another ladder leading to the forward entrance of the wardroom. What they saw next would be puzzling.[26]

Whereas the ammunition for the 5-inch 38 twin mounts were stowed below and brought up by mechanical hoist on an as-needed basis, the 20mm gun mounts had small ready service rooms otherwise called "clipping rooms"—essentially big steel boxes—located nearby that contained multiple rounds of the small caliber shells. One of these steel boxes, tucked in between mount three and the bridge, took a direct small-caliber hit, resulting in the progressive burning of the aluminum as cartridges cooked off, bursting open the box that sent molten metal flying and set adjacent boxes ablaze.

Because *Atlanta*'s topside superstructure had been fabricated mostly from aluminum to prevent the ship from being top-heavy, the intense heat from the popping off rounds burned holes into the deck which the box sat on, dropping the fire and additional rounds of cooking ammunition into Captain Jenkins's and Rear Admiral Scott's staterooms. As the decks for the sleeping quarters of *Atlanta*'s two most senior officers were also fabricated with aluminum, this deck also melted away, dropping flaring ammo into the wardroom. Not able to burn through the steel main deck, the cartridges glowed like fireflies, occasionally cooking off and releasing the shells.[27]

As McKinney stepped through the wardroom not quite figuring out what was causing the firefly phenomena, he stumbled over a knocked-over chair. Picking himself up, he and Seaman 2/c Curtin opened the watertight doors leading to the exterior main deck on the starboard side, ripped off their masks, gulped in fresh air, and proceeded to release their heated chemical canisters over the side. McKinney recalled a shipmate running by yelling "Abandon ship ... she's going to blow!" McKinney and Curtin considered their options. The waters in Ironbottom Sound were known to be shark infested. Given a choice of being eaten or blown up, the two chose the latter ... and then proceeded to act to prevent the latter from happening. Noting that a stack of RBA canisters had been stowed outside of the wardroom, McKinney and Curtin each grabbed a canister, reactivated their breathing rigs, passed back through the wardroom to head back to their GQ station. Back in their berthing space, they checked the fire main running through the compartment to discover the pressure seemed adequate. They then grabbed a firehose and proceeded to haul it topside.[28]

A similar situation occurred on the port side abaft the bridge when a hot fragment of an exploding incoming round penetrated a clipping room that stored 1.1-inch rounds for an AA station located one level over the main deck. Again, *Atlanta*'s sailors had to respond to a situation amidships where ammunition was cooking off.

Another enemy shell hit a pyrotechnic locker in the forward superstructure, setting off yet another bright blaze that proved to be a challenge to combat. Furthermore, all three forward 5-inch 38 mounts were knocked out of commission, thanks to ordnance dinging the barrels; the steel boxes shielding the guns being penetrated; and hits that penetrated the armored bulkheads of upper handling rooms. Those sailors who escaped from the damaged forward twin mounts were subsequently cut down by the Japanese bombardment shells that fragmented upon detonation. The number of fatalities up forward was staggering.[29]

Regaining consciousness, Shaw staggered out on to the quarterdeck. "I tripped over two of the lads I'd sent ahead, dead on the deck," he would later tell his son. Shaw sagged back against a bulkhead, fighting off nausea and holding his hand against his chest to counter the intense pain.[30]

Adrift, the crippled light cruiser suddenly came under fire from *San Francisco* positioned a mere 2,000 yards abaft *Atlanta*'s port beam. Two or three 8-inch salvos from the heavy cruiser's main battery—again thanks to smokeless gunpowder—lit up the ship before Rear Admiral Callaghan reportedly ordered "Cease Firing, Friendly ship." However, the heavier caliber armor-piercing shells had done their damage, punching into and out of the opposite side of the aluminum forward superstructure beneath the pilothouse where Rear Admiral Scott, Captain Jenkins, and the rest of the bridge watch team were stationed.

One round apparently came in higher, piercing the modest 4-inch steel armor of the pilothouse aft on the port side. Diverted forward, the shell exited out the starboard side with catastrophic effect. Captain Jenkins would later recall that he and Rear Admiral Scott had moved over to the starboard bridge wing to observe the course of the sea battle that had moved northwards when something occurred on the port side to cause Jenkins to investigate. Upon returning following a blast, he was stunned to see no starboard bridge wing. Mustin later lamented: "A couple of his staff officers were standing behind him and they simply vanished. There was no burial at sea—there was nothing to bury. They were just gone."[31]

The shell also proved fatal to *Atlanta*'s Navigator, Lt. Cdr. James S. Smith and the communications officer who served as the Battle Station Officer of the Deck, Lt. Cdr. Philip T. Smith, Jr. Jenkins did not get away unscathed either as he suffered a foot injury that impeded his mobility. He eventually made it back to the aft conning station at Battle II, about fifteen minutes after *Atlanta* had opened fire, to join with the XO, Mustin, and other surviving officers. The protective armor of the pilothouse along with his life jacket and helmet saved Seaman 1/c Reed who stood just outside. Flying metal sheared off the top of the steel pot covering Reed's cranium and cut into his vest. One piece of shrapnel did find his neck. The wounded sailor scurried to the port ladder leading below, but fell two decks and broke his arm upon discovering the ladder was missing.

Another topside survivor, Ensign Graff, had been knocked to his knees when the torpedo hit. Subsequently burned from incendiary material, he quickly concluded following the *San Francisco* salvo that the signal bridge on the port side was no place to be: "All the ladders had been shot away." He crawled through the pilothouse over the bodies of everyone who had been killed and out to the starboard side. Recalling the experience years later, he crawled down to the gun mount below and then fell to the main deck where Ed Corboy found him unconscious.[32]

The Assistant First Lieutenant, Lt. Van Ostrand Perkins, was stationed at Repair III amidships aft of the bridge. The impact of one of *San Francisco*'s incoming shells shattered his kneecap, with flying shrapnel causing what would look like a sword cut from the middle of his lower lip down his chin and neck. Fireman 1/c Charles Dodd and surviving shipmates who had escaped from hits against the 1.1-inch gun mounts, lowered themselves on to the 0-1 level on the starboard side where they saw Perkins fashioning a crutch out of a swab.[33]

At about the time that the fatal blow occurred to the bridge, another grouping of armor-piercing shells passed through the aft superstructure at the main deck level below where Commander Emory was standing in Battle II, and took out the port and starboard 5-inch 38 mounts. As the mount on the port side had been trained forward, it received two 8-inch rounds that entered on its left side. With one of the shells hitting the breech mechanism, the resultant explosion blew out the back of the gun mount with fatal effect for those inside. Other rounds simply punched through the aluminum superstructure to damage the corresponding starboard 5-inch 38 mount. As if there was any doubt as to the culprit in this regrettable friendly fire incident, the after group of shells left ample forensic evidence as armored piercing rounds at the time were loaded with a very fine dye powder at the tip of a collapsible nose cone that would disperse upon water entry and color the subsequent splash. Ships were assigned different color dyes. *San Francisco*'s was green.

Mercifully for *Atlanta*, the northerly direction of the American column and the decision of the Japanese forces to also turn northward took the light cruiser—illuminated by blazes on the forward section of the ship—out of the battle.

Boatswain's Mate Spadone climbed down to the quarterdeck area after the gunfire had ceased and came upon Lieutenant Shaw just inside a hatch leading towards the engineering spaces and asked the dazed officer for instructions. Shaw focused on the situation and told Spadone to undo life rafts that were lashed along various exterior bulkheads and lower them into the water with wounded sailors. With *Atlanta* listing to port and with blazes forward, Shaw's guidance seemed prudent as there were grave concerns that the forward magazine could blow. Back aft additional rafts were placed in the water with some of the wounded. Told by his mount captain to abandon ship, Dunaway was one of those who rode on one of the rafts for the rest of that night. Seaman 1/c Alvin Otto, who came up from below where he had been stationed in the ammunition handling room for the decimated mount five on the starboard side, recalled getting off and on the boats throughout the nights, making runs to get medical supplies to care for the wounded. One of those wounded sailors was Seaman 1/c Reed who recalled regaining consciousness in a main deck passageway to find that someone had thoughtfully attempted to splint his broken arm with a monkey wrench. Reed got up and walked out onto the exterior starboard passageway and headed aft where he was placed into one of the rafts.[34]

With the fire mains piping rendered useless throughout the forward section of the ship thanks to no water pressure, crewmembers formed bucket brigades to get water onto the flames. Lacking rope, the crew tied long battle telephone lines on bucket handles to lower buckets overboard to fill with seawater. Also to dampen down some of the blazes, damage control teams used portable gasoline-powered pumps that sucked up water into a hose dropped over the side. Other remaining able crewmembers contended with other fire and flooding issues which had been thrust upon the ship. Despite having removed flammable linoleum from the decks and scraped paint off the interior bulkheads weeks before, none of the directives had mentioned an obvious hazard that existed in compartments throughout the ship—tech manuals for all of the pieces of equipment as well as personal libraries full of quantities of reading materials. All were kept on these open shelves that had a movable bar in front to keep the books in place during heavy seas. *Atlanta*'s firefighters discovered such paper materials, not stowed in a flame retardant box or cabinet, were very combustible.

Many shells that penetrated the forward superstructure entered living and workspaces. Two examples included the gunnery office and first lieutenant's office where small caliber shells had detonated and caused "astonishing damage." Joiner bulkheads—the thin partitions that divided offices, staterooms, and passageways—were blown down. Burning mattresses tended to smolder, creating a thick acrid smoke.

The ever-eloquent Morison portrayed the main deck as a charnel house—"Burned and eviscerated corpses, severed limbs, and chunks of flesh mixed with steel debris, littered it from stem to stern." He went on, "Blood, oil, and sea water made a nasty, slippery slush through which one could only move on hands and knees."[35] If moving horizontally through the ship posed a challenge, vertical movement proved more daunting, especially since exploding Japanese and American ordnance had damaged or stripped off exterior ladders from the forward superstructure. To get at the forward 20mm clipping room fire, exhausted sailors scaled up the sides of the deckhouse that had been spattered with oil and water, avoiding loose rolling shell casings and stepping around dead bodies. Led by Lieutenant McEntee, who had climbed back aboard, the damage control team worked a hose up the less-damaged starboard side of the ship and extinguished the blaze. In doing so he burned both of his hands.[36]

Other wounded officers fought off the pain of their injuries to assist with the firefighting efforts. After supervising an effort to lower wounded into a life raft, Lieutenant Shaw headed aft, working his way through debris and smoke to see if he could get his injured hand tended to. Finding the ship's doctor taking a brief break after working to stabilize several severely injured crew members, Shaw, believing he had several fractured bones, asked if his hand could be splinted. Lieutenant Commander Garver looked at him, "Stuff it in your life jacket, I haven't time to care for it now." Ignoring the pain of his injuries, Shaw headed back towards the flames amidships to see what could be done. Meanwhile his good friend Lieutenant Perkins, having organized a firefighting party to attack one of the blazes, struggled using his swab crutch to head aft over a deck covered with oil, blood, as well as smashed potatoes from a burst open spud locker. Touching his chin, he felt his oozing blood and recognized he had an additional wound to contend with. He then looked up to see Jim Shaw. Carrying morphine syringes, Shaw used his weaker left hand to inject Perkins with morphine. "Son of a gun, that hurt more than my leg," was probably a censored version of what Perkins actually said. Perkins, in turn, gave his buddy a morphine shot to ease the pain of his smashed hand.[37]

With the pain-suppression medication applied, the two officers paired to lead additional firefighting efforts with Perkins acting as an on-scene leader and Shaw serving as a runner to pass along progress reports to Battle II where Captain Jenkins oversaw the effort to save his ship.[38]

The aforementioned blaze in the pyrotechnic locker located along the port side of the forward superstructure generated such heat as to cut into the base of the foremast which had been constructed of aluminum. Sheared off, the mast with its air search radar antenna toppled over to one side of the ship. Mustin located one of the crewmembers who knew how to use an oxyacetylene torch to try to cut the mast free, but the welder made little headway before running low on gas. There were other problems to deal with.[39]

To correct a port list that flooding caused by the torpedo hits and the toppled mast had created, Lieutenant Commander Nickelson released the port anchor and the chain.[40] McKinney, sharing concerns of his shipmates that the forward magazine was still in danger, approached Nickelson with the idea of flooding his former berthing space. Nickelson had no objections but suggested he check with the XO. McKinney made his way back to the aft section of the ship where he found Commander Emory in his stateroom beginning to compile material for the post-action reports: "He ok'd my plan but urged me not to sink the ship." McKinney then returned below to flood his space, covering the deck with a buffer of about a half foot of water.[41]

With *Atlanta* drifting with the winds and currents, as a precautionary measure against grounding up against the Japanese-held side of Guadalcanal, Lieutenant Commander Nickelson led a groups of sailors who struggled to lower the starboard anchor chain some 100 fathoms, with the hope that if the anchor could catch the bottom before the ship beached too close to the shore.

As the anchor detail worked up towards the bow, every so often there were indications of a large vessel passing to starboard. With *Atlanta*'s rated torpedomen ranks decimated during combat, volunteers rushed to the starboard tubes, anxious to take a revenge shot at an enemy warship. Mustin, a bit cautious, grabbed Signalman Striker John W. Harvey who, because of his Battle II GQ station, had survived the battle unscathed. Harvey peered through his binoculars into the darkness and as the mystery vessel emerged, he exclaimed "It's the *Portland*!" Captain Jenkins responded, "Are you sure?" Harvey indicated that he could make out the heavy cruiser's distinct clipper bow, forward tripod mast, and the separation between the forward and aft stacks. Convinced of the vessel's identity, Jenkins instructed Harvey, "Very well then, see if you can signal her."[42]

Harvey picked up a handheld signal lamp and flashed over. Receiving a quick response, Harvey signaled it was *Atlanta*. However, the young striker had difficulty discerning the next flashing light transmission and when he asked *Portland*'s to repeat, *Portland* responded asking for transmission of that day's recognition signal. Panic ensued in that the code book dictating those signals had been kept up on the now-decimated signal bridge. Fortunately, as Harvey attempted to signal that no code book could be located, the first glimmer of light in the east made it easier for *Portland*'s lookouts to identify *Atlanta*'s silhouette despite the missing forward mast.[43]

As daylight appeared, the two battered warships gave testament to the ferocity of the previous evening's slugfest. Within the waters of Ironbottom Sound that morning, besides the two cruisers, the destroyers *Cushing* and *Monssen* remained afloat, aflame and abandoned, destined to add their hulls to the growing fleet resting on the seabed below. In the distance, *Aaron Ward* floated motionless, surviving without propulsion despite having withstood blows from 14-inch battleship shells and multiple smaller caliber blows.

Atlanta's men could not see *San Francisco* nor most of the other American ships that had entered the fray hours earlier. Three Japanese warships had taken advantage of Callaghan's ceasefire order to savage *San Francisco*'s superstructure. Their fire had killed not only the task force commander but also the ship's captain, Capt. Cassin Young, and nearly everyone else who had been stationed on the bridge. The heavily damaged cruiser would join with five other American combatants to extricate themselves from the waters off Lunga Point and steam south through the Indispensable Strait with the objective of reaching the sheltered waters of Espiritu Santo. Besides *San Francisco*, the six-ship formation included destroyers *O'Bannon*, *Sterett*, *Fletcher*, and cruisers *San Francisco*, *Juneau*, and *Helena*. Capt. Gilbert Hoover, *Helena*'s commanding officer, assumed command as the senior surviving U.S. naval officer of these escaping remnants of the task group. Of the six ships, only *Fletcher* steamed unscathed. *Atlanta*'s twin *Juneau* also had taken a torpedo on her port side a bit further forward, eliminating her forward fire room. With intact aft engineering spaces, she was able to continue on. *Atlanta*'s crew would later learn that the destroyers *Laffey* and *Barton* had been lost during the battle.[44]

The Japanese also suffered losses. Thanks in part to *Atlanta*'s quick reaction, the Japanese destroyer *Akatsuki* was no more. Seaman Harvey peered out with his binoculars to view a vessel "with what we called an 'East Coast paint job' i.e., zebra stripes of camouflage blue and gray." Noting steam coming from its forward stack, Harvey further studied the stack arrangement and location of gun mount to conclude she was Japanese. The young signalman striker was correct. Drifting between Cape Esperance and Savo Island, the crippled destroyer *Yudachi* had no power. *Portland* took aim at the wounded enemy warship with her forward 8-inch triple turrets. On *Atlanta*, the sleep-deprived sailors who were struggling to salvage the ship took a respite to watch *Portland* fire multiple salvos, straddling *Yudachi*. Petty Officer McKinney recalled that he and his shipmates stood along the lifelines, "... spectators to an action rarely witnessed." Finally, a salvo hit the enemy tin can amidships detonating a magazine, resulting in a mushroom cloud that captured the rays of the just-risen sun. The *Yudachi* rolled over, exposing her copper-colored bottom which elicited a cheer from *Atlanta*'s white hats. McKinney recalled one shipmate did not share in the joy, observing: "Don't cheer fellows, the poor guys are dead. That could have been you."[45]

One more Japanese warship lay on the horizon to the northeast of Savo Island. The battleship *Hiei* had been hit hard, particularly by close-range salvos from *San Francisco*'s 8-inch batteries, and had suffered blazes throughout her topside areas. However, her watertight integrity was maintained, except for a hole in her starboard quarter that led to flooding in her after steering compartment which shorted out the steering engine.

Spying the crippled *Aaron Ward* in the vicinity close to Florida Island, the Japanese battleship sought to avenge the loss of *Yudachi*. However, as the third and fourth 14-inch salvos straddled the American tin can, the cavalry came to the rescue in the form of Marine SBD Dauntlesses from Henderson Field. The Cactus Air Force had enjoyed an uneventful night thanks to the gallantry exhibited by the ships of Task Group 67.4 earlier that morning. At just after 0700, *Hiei* absorbed a 1,000-pound bomb which diverted attention away from *Aaron Ward*. The crippled destroyer also received additional salvation when the fleet tug *Bobolink* arrived to tow her to the sanctuary of Florida Island. The tug then proceeded to move *Portland* to a safer haven. Meanwhile, *Hiei* endured repeated air attacks throughout the day, not only from the Cactus Air Force but also from *Enterprise*. A torpedo plane attack in the afternoon punched two holes in the starboard side that led to pervasive flooding, listing the Japanese dreadnought to starboard and down by the stern. That night *Hiei* was abandoned and would sink north of Savo Island.[46]

For the survivors in *Atlanta*, the distant struggle of the *Hiei* and the relay of aircraft coming and going from Henderson Field were mere distractions from the efforts to care for the wounded and attempt to save the ship. Upon learning that *Portland* had suffered a torpedo hit that jammed her rudder and forced her to steam in clockwise circles on her port engine, Jenkins had a thought. Since *Atlanta*'s aft emergency diesel generator was providing power for rudder control, perhaps the two ships could latch together and depart the dangerous waters using *Portland*'s port engine and *Atlanta*'s rudder? Apparently after Jenkins floated the idea with his counterpart, Capt. Laurance DuBose, in *Portland*, the two concluded the challenge of maneuvering the ships alongside appeared insurmountable. However, *Portland* assisted in other ways. Having line-of-sight communications with Lunga Point, the heavy cruiser relayed Jenkins's requests for small craft to evacuate the wounded.[47]

Noting the currents pushing the ship towards the Japanese-held section of Guadalcanal some 3 miles distant, Lieutenant Commander Mustin managed to open the armory after locating the key on a dead gunner's mate and issued Springfield rifles and ammunition, keeping a rifle for himself. It was a decision he later had regrets about as some of the crew, seeing Japanese survivors in the water, began firing on them. Orders passed to cease fire. In the meantime, small boats arrived from Guadalcanal to take off *Atlanta*'s wounded and recover survivors in the water. *Atlanta*'s motor whaleboat contributed to the search effort. As the boats approached Japanese survivors, most swam away to avoid capture—not all. When one of the Cactus Navy landing craft pulled alongside, *Atlanta*'s sailors looked down to see a Japanese boatswain's mate bedecked in his jumper white uniform tending to some half dozen wounded oil-covered survivors of both nationalities who had been scooped up en route. Standing up, the boatswain gestured for some rags to clean off the men he had taken under care. His request was met.[48]

As the morning progressed, water continued to seep in, creating a discernible list to port. To lower the ship's metacentric height, the point at which a ship could tip over, Jenkins's crew worked to lighten the topside weight. With the port 5-inch 38 gun mount ripped to pieces, hunks of shredded steel were heaved overboard. Ammunition that had been stowed in the handling rooms found its way over the side. *Atlanta*'s able-bodied survivors struggled to free the shot-up boats and cut the mast that hung over the side with little success. As the port side torpedo tubes were trained out, the fish were released overboard to further lighten the load. All but three 300-pound depth charges went over the side after they were disarmed in addition to the smoke canisters that Mustin had wanted to activate during the battle.[49]

In cases where bodies were severely mutilated, the remains were lowered over the side. For those sailors who were killed but remains were relatively intact, *Atlanta*'s aft berthing spaces were pressed into service as a morgue. Electrician's Mates Bob Tyler and Bill McKinney helped clear the decks of the carnage. Coming upon a large, deceased boatswain's mate, the two attempted to lift the shipmate alone only to have him come apart at the middle. Each half was lowered overboard. McKinney came upon a liberty running mate on the starboard quarterdeck who was missing a leg and had a jagged piece of metal embedded in his chest. Responding to a weakly made request for water, McKinney provided his friend a sip only to have him vomit up blood and then die. McKinney then came upon another shipmate lying exposed to the morning's sun. McKinney thought to cover his face with a clean handkerchief, but the young man tore it off, fearing that he would be thought dead and tossed over the side.[50]

The battle would eventually claim 170 of *Atlanta*'s crew of 735. Commander Cooper, Lieutenant Commander Garver, the dentist Lieutenant (jg) Erdman, pharmacist's mates, and other shipmates performed heroic service in their care for those who were wounded.[51]

In addition to the two port and starboard mounts that had been taken out of service due to *San Francisco*'s misdirected gunfire, Japanese rounds took out the mount directly behind of the aft gun director leaving *Atlanta* just two unscathed twin mounts near the stern section of the ship. Signalman Striker Harvey came down from Battle II to eat some fruit that was the limit of *Atlanta*'s food offerings, and attempted to walk forward on the starboard side—a challenge due to oil and debris. He noted the starboard motor whaleboat was being lowered but not much could be done with the remaining small boats that were heavily splintered. Peeking into the burned-out wardroom he noted the hole in the overhead and was struck by the remains of a stand-up piano. Harvey walked up further and looked up to observe how freakish mount three looked:

> The trajectory of the falling projectile cut one barrel completely off and left its partner hanging by a mere thread. An explosion, or backfire in this turret caused by the damaged barrels caused numerous casualties. #2 Gun was O.K., however, the handling room directly underneath, on main deck forward of the wardroom was a wreck. #1 turret had a 5" hole directly through the trainer's seat. With an absence of a hole in the back of the turret, the projectile must have gone off in the turret. All three forward guns were trained just about 10 to 15 degrees to port when loss of power occurred.[52]

Working his way back aft on the port side, Harvey observed the devastation wrought on the AA batteries but also found the presence of colored threads baffling until he looked up to see the ruined signal bridge and shot-up flag bags.

The concern of drifting within range of small arms fire from Guadalcanal was alleviated when *Bobolink* appeared. McKinney and others helped to lay out the towing cable that would be attached to the vessel which had been built during World War I as a minesweeper and had been recently converted to a fleet tug. The dark green-painted vessel took *Atlanta* under tow with her anchor dragging some 600 feet below. To help with the movement, Captain Jenkins used a megaphone to shout steering orders down to the main deck where they were relayed aft to the hatch leading to the after steering compartment. After two hours, as the damaged warship approached the safe waters off Lunga, the hook touched bottom.[53]

While under tow, the aft generator also powered an emergency radio which kept Captain Jenkins attuned to the tactical situation. Suddenly, there came a report of incoming enemy aircraft. Could the aft diesel generator power the aft mounts? The gun mounts fed off an electrical switchboard in the upper level of the partially flooded after engine room. What *Atlanta*'s electricians needed to do was run a cable up the escape trunk, back 60 feet, and then down the trunk leading to the diesel generator. Through trial and error the electricians succeeded in the nick of time.[54] Over the horizon came a Betty bomber which flew toward *Atlanta*. The crippled light cruiser responded. The Aft Director Officer Lieutenant (jg) Newhall followed the incoming Betty optically and called down the sight settings and fuze settings to the mount, and with the order to open fire, loud claps from muzzle blasts cut through the morning air. The Betty turned away unscathed.[55]

The exhausted able-bodied crewmen went back to work trying to save the ship. A few fires remained to be extinguished including one in Lt. Bill Mack's stateroom that was being fed by his rather large collection of vinyl recordings. Armed with a CO_2 canister, Chief Shipfitter Emmitt Grounds donned an RBA, entered the smoke-clogged compartment and stumbled on the remains of Metalsmith 1/c "Pappy" Bremmer, who had been among the more elderly enlisted, as well as a felled electrician's mate and radioman.[56]

At 1101, *Atlanta* became the sole surviving ship of the two light cruisers built at Federal Shipbuilding the previous autumn when a torpedo from Japanese submarine *I-26* struck *Juneau*. Lt. Cdr. Bruce McCandless, one of the surviving senior officers remaining in *San Francisco* observed: "The *Juneau* didn't sink—she blew up with all of the fury of an erupting volcano." With pieces of what was once the *Juneau* showering down on the remaining ships of the flotilla, the sudden shock of the instantaneous loss of this American warship and her crew of some 700 sailors confronted Capt. Gilbert Hoover, as the officer in tactical command, in *Helena*. He faced the dilemma of diverting one of his ships back to attempt a recovery operation for survivors, knowing that a Japanese submarine lay below awaiting an opportunity to sink yet another American warship. With him, he had the damaged cruisers *Helena* and *San Francisco*, both carrying sailors who needed medical attention, as well as the damaged *Sterett* and the unscathed *Fletcher*. Earlier, Hoover had detached *O'Bannon* to transmit news of the early morning fight to Halsey's headquarters in Nouméa, with the thought that enemy radio detection finders would not zero in on the location of the crippled formation.

Rather than risk further attack, Hoover decided to press on to place more distance between him and the Japanese submarine. Twenty minutes after the torpedo attack, a B-17 Flying Fortress that had been requested to provide air cover over the formation, appeared overhead. Rather than break radio silence, one of *Helena*'s radiomen sent a flashing light signal to the bomber to inform of the *Juneau*'s loss and where she had been hit, and to inform Halsey that survivors could be in the water. Indeed there were. Some 120 men, some badly injured, bobbed in the pitiless waters hoping for a rescue that did not materialize. Many were able to climb on to rafts and life nets. With San Cristobal visible on the horizon, the survivors futilely attempted to paddle there, including George Sullivan who lost his four brothers in the torpedo blast. Air search efforts to locate the survivors proved inept as hours stretched into days and the sun, lack of water, and finally aggressive sharks whittled down the number who survived the ordeal to 10. George Sullivan was not among them.[57]

At Espiritu Santo, Halsey's subordinate admirals stationed there interrogated Hoover about the circumstances surrounding the loss of *Juneau* and the decision not to risk a recovery, and efforts—that proved inadequate—to inform senior leadership of the light cruiser's loss. They made a recommendation to him that Hoover be relieved of his duties in command of *Helena* which Halsey approved.[58]

Back on *Atlanta*, unaware of the fate of their sister ship, the question was could the light cruiser be also spared a watery grave? Thinking they could pump out the after fire room, McKinney and two of his shipmates set up a bilge pump outside the hatch leading into the space, only to have water squirt out under high pressure as they attempted to undog the hatch.[59] Captain Jenkins conferred with his senior officers. Several key officers in the wardroom had been killed, including the chief engineer, the navigator, and the communications officer. Both the First Lieutenant/Damage Control Officer, Lieutenant Commander Sears, and his assistant, Lieutenant Perkins, had been wounded and been taken ashore. The gruesomeness of the toll taken fell upon Electrician's Mate 1/c Bob Tyler who dared work his way up through the bridge. On the starboard bridge chair, the lifeless Navigator, Lieutenant Commander Smith, sat unscathed. Otherwise there was a headless body slumped over against a screen; at the helm were the legless remains of the chief quartermaster; on the deck was a hand with a Naval Academy ring "Class of 1911" imprinted—Rear Admiral Scott's graduation year. Looking up on the bulkhead, the pilothouse clock read 0215. Tyler peered a deck below into what was Radio Central where explosion and subsequent flames essentially cremated the approximately dozen radiomen who were on duty during the hostilities.[60]

The remaining senior leadership included the CO, XO, Assistant Engineering Officer Lt. Cdr. Jack Wulff, the Gunnery Officer Lt. Cdr. Bill Nickelson, and Lt. Cdr. Lloyd Mustin. Jenkins designated Nickelson as the new damage control officer.

Inspecting the superstructure and the hull, Mustin counted 53 different shell hits.

With the freeboard on both sides of the ship decreasing, the obvious deduction was that water was still seeping in. Wulff and Mustin believed that if the ship could be shifted to a sheltered anchorage such as Tulagi Harbor or just off Lunga, it could be possible to plug the voids between the aft engine room and the forward fire room; dewater the forward fire room; and get the ship underway using the turbines located in the undamaged forward engine room. There was concern about an unexploded torpedo that was discovered planted in the port side. Would it suddenly detonate? Mustin felt they were safe in the belief that the exploders never had enough run time to arm.

The new damage control officer and XO felt differently. Accounting for the exhausted crew and the vulnerability of the ship, they recommended scuttling the ship. Captain Jenkins concurred and radioed to Captain DuBose over in the *Portland*, as the senior officer present afloat, to request permission to scuttle his ship. DuBose granted the request. Earlier, he had offered to sink *Atlanta* with his gun batteries if required but with his ship now under tow by *Bobolink* to Tulagi, the offer had been rescinded. Meanwhile, many of the 245 able-bodied sailors who remained aboard following the initial evacuation of severely wounded shipmates also required medical treatment. Included among those who were able-bodied were Lieutenant McEntee, who had both hands wrapped in bandages following his leading the charge to put out the forward clipping room blaze.[61]

With the decision to scuttle the ship, Guadalcanal sent additional boats to off-load the crew. Seacocks and hatches throughout the ship were opened with one exception. The hatch leading to the aft crew berthing space was sealed as the temporary morgue was to become a crypt.[62] Gunner's Mate 2/c Joseph Allen led a small party down into one of the magazines and retrieved a box marked "Demolition Gear." A demolition charge was placed on a bulkhead in the aft diesel engine room adjoining one of the aft 5-inch magazines. Meanwhile those having access to their belongings packed bags and pillowcases with letters, pictures, toiletries, skivvies, and other uniform items. McKinney returned to his partially flooded berthing space to hastily fill a pillowcase with shoes, socks, uniform items, and a box of cigars.[63]

With the crew departed, two boats awaited the scuttling party of 10 led by Captain Jenkins. Lieutenant Commander Wulff planted the demolition charges in the steering compartment adjacent to the aft magazines. At about 1730, Jenkins, being last to depart, flipped a switch setting off the explosives placed in the stern and climbed down a cargo net into an awaiting motor whaleboat. The two boats sped away and stood off a few hundred feet—a safe distance for the occupants to watch the light cruiser go down.

Heavy smoke billowed from the stern. On occasion, a "whump" could be heard from the aft magazine; however, when Jenkins ordered his coxswain run the motor whaleboat up to the bow to see if the ship was settling, it became readily apparent that *Atlanta* intended to stay on the surface. Further annoying Captain Jenkins, it started to rain. Jenkins thus ordered the second motor whaleboat, with Lt. (jg) John P. Luddy, to speed out to *Portland* to see if her guns could be used to deliver the final blow. Upon reaching the heavy cruiser under tow by *Bobolink*, Captain DuBose communicated back that he could not spare the time, as he sought to shelter his ship at Tulagi that evening.

After daylight faded all that could be seen was a large blaze engulfing the stern that finally triggered an explosive chain reaction. Signalman Striker Harvey later wrote: "… there was a great explosion at 2015 and a burst of flame, and then nothing." Jenkins ordered the coxswain to steer the boat to the site of the blast only to find debris, oil, and the rudderless motor whaleboat that had been tied alongside. Convinced his former command was no more, Jenkins directed the coxswain to steer for Lunga Point.[64]

CHAPTER 13

Homecoming

With the steady flow of survivors coming ashore all day from the American warships lost off Guadalcanal that previous evening, the Navy contingent ashore set up a station to capture the names of survivors. Petty Officer McKinney remembered having to state his name, rate, and serial number to an awaiting scribe. Those who needed care for wounds were taken to medical facilities while others were directed to campgrounds. Seaman 1/c Reed, riddled with shrapnel, found himself under the care of a Navy doctor who also hailed from Mystic, Connecticut. Chatting, Reed learned the fellow lived next to one of his older brothers. For the young sailor the hometown connection provided a confidence boost that all would turn out well.[1]

Directed to a coconut grove, a contingent of *Atlanta* survivors arrived at a designated camping site where some Marines pointed to nearby foxholes where emergency rations had been stashed. Tents were provided and the sailors took to the challenge of figuring out how to erect them. Soon a little tent village emerged.[2]

McKinney and a few others remained at the water's edge. Filthy, the electrician's mate dispensed with his ragged dungarees and torn-up shoes, grabbed a bar of soap and cautiously waded out to take a saltwater bath. Having cleaned himself up he then grabbed a pillowcase with uniform items and a box of cigars that he had hastily packed in darkness before he flooded out his former berthing space. To his horror, all he had for clothes were a pair of shoes, socks, and three white jumpers—no pants or skivvies. With darkness descending and wearing just a white jumper with shoes and socks, McKinney headed off to where he thought the *Atlanta* camp had been pitched but suddenly found himself maneuvering over strands of barbed wire, clutching his box of cigars with one arm and protecting his genitals with the other. McKinney felt something whiz by his head and then another. A voice yelled, "Get down you damned fool, that's a sniper!" Down on his hands and knees, as the half-naked sailor crawled towards the direction of the voice he heard another, "With that nice white jumper and big red ass, how did they ever miss him?" With that sarcastic beacon McKinney managed to reach the Marine outpost. Redirected, McKinney arrived at the *Atlanta* campground late and was given a tent to erect. Too tired, he and another sailor simply chose to lie on the ground and pull the canvas over them. A heavy downpour turned the ground to mud which only slightly disrupted McKinney's slumber.[3]

As night fell, rations from the foxholes were gathered for the evening meal. Jenkins was a late arrival as his motor whaleboat arrived at the Point Lunga boat dock well after dark. He and other officers of the demolition party headed off to the main Navy camp located a short distance away from where most of *Atlanta*'s refugees had taken cover. There, Jenkins and Emory spent the evening hosted by Capt. William G. Greenman. Greenman, the former commander of the cruiser *Astoria*, was reassigned to run the small Navy shore establishment on Guadalcanal and Tulagi following the loss of his ship at the Battle of Savo Island. Greenman's chaplain made room for some of the *Atlanta*'s junior officers in his tent. Lieutenant (jg) Corboy recalled being assured by a staff officer that they were in for a quiet night: "Get a good sleep and forget all about it." Corboy remembered being "asleep almost before we hit our cots."[4] Back on the beach, an old Japanese Ford truck picked up Signalman Striker Harvey and other enlisted shipmates from the two demolition party boats and dropped them off after 2200 at the *Atlanta* tent farm. Harvey found an empty space on the floor and using his lifejacket as a pillow, soon fell asleep.[5]

With the Japanese heavy cruisers *Suzuya* and *Maya*, and several Japanese destroyers appearing offshore 90 minutes after midnight, the exhausted *Atlanta* bluejackets would be in for a rude awakening. With the sky illuminated by flares dropped by a Japanese floatplane, the enemy warships commenced a 31-minute barrage of Henderson Field. Over that period, the two heavy cruisers alone fired 989 8-inch projectiles that would rumble the earth around the newly established survivor camps.[6]

The concussion from the broadsides, the whistling of shells through the treetops over their tents, and the crashing of the shells nearby caused the *Atlanta* survivors to scamper from their tents to the nearby foxholes. The explosions woke up McKinney who observed men shouting and running in all directions. Harvey noted many in his tent jumped and bolted for air raid shelters, clueless as to where they were located. Others leaped into foxholes. With a realization after a few minutes that the Japanese were targeting the airfield, several left the safety of their foxholes to get a better view.[7] McKinney made out the Japanese ships as they paraded back and forth: "I could see the little winking pinpoints of blue light as their salvos thundered toward us." McKinney later wrote that it felt like "lying across several railroad tracks with a fast train approaching on each one." One petrified sailor who yelled "This one is for us!" with every incoming salvo was subdued with a fist to the face. Over at the field hospital, Seaman 1/c Reed recalled being brought into a large dugout for nominal protection against the exploding shells.[8]

With the Japanese flotilla steaming away, the rattled *Atlanta* shipmates meandered to their tents to attempt a few more hours of rest. Unfortunately for a few *Atlanta* sailors who dove into a latrine pit instead of a foxhole, they would be banished, which enabled McKinney to sleep the rest of the night under cover. The next morning, the damage assessment determined that most of the Japanese shells landed around the auxiliary airfield Fighter One. The bombardment claimed two Wildcats, an SBD, and damaged another 15 fighters. In contrast, Henderson Field remained relatively unscathed. The Japanese warships that delivered the blows would not. The slumbering sailors were likely awakened by aircraft engine sounds as five dive-bombers and three torpedo planes escorted by 10 fighters left Henderson to run down the Japanese bombardment force. *Enterprise*, operating some two hundred miles southwest of Guadalcanal, also launched scout bombers followed by a strike group. One scout bomber scored a hit forward of the bridge of the cruiser *Kinugasa* claiming the lives of the CO and XO. Another SBD, nicked by antiaircraft fire, plunged into the *Maya* causing considerable damage. Seventeen of *Enterprise*'s Dauntlesses finished off the *Kinugasa* and damaged the cruisers *Chokai* and *Izuzu*.[9]

Back on Cactus, a long chow line wound through coconut trees as refugees from *Atlanta*, *Monssen*, *Cushing*, and *Barton* waited for their first meal of substance in a day and a half. A Marine sergeant spotted McKinney, wearing his white jumper and shoes and socks, in the line and pulled him aside. After a few minutes, McKinney was back in the line wearing the spare uniform of a Marine who had been killed the day before. Using a coconut shell as a mess kit, McKinney received a ration of corned beef and captured Japanese rice. "Delicious!" he later wrote. After the meal, Signalman Striker Harvey recalled the *Atlanta* sailors were told to break down their tents for consolidation at a location nearer to Lunga Point.[10]

At this point, *Atlanta*'s veterans were seemingly seated on the sidelines as the ongoing struggle for Guadalcanal approached a climax. However, an opportunity to get back into the fight was about to present itself. The valiant effort by Callaghan's 13-ship flotilla had merely disrupted the Japanese timetable to deliver transports full of Imperial Army troops and supplies by 24 hours. With the cruiser bombardment that aimed to suppress the Cactus Air Force, a 23-ship convoy centered around 11 transports commanded by Rear Adm. Raizo Tanaka was now en route from Shortland Island.[11]

Aiming to arrive at Guadalcanal late that evening, Tanaka's convoy steamed under clear skies that morning. That changed at 1250 as 18 SBDs and seven TBFs from Henderson Field attacked one of three of Tanaka's transport formations, sinking two *Marus* and sending a third scurrying back to Shortland. Additional attacks from Henderson and *Enterprise* sank four more *Marus*. As Japanese destroyers maneuvered to pick up soldiers off the sinking transports, Tanaka ordered his remaining four transports to continue, guarded by five destroyers with four additional destroyers catching up following the recovery effort. As darkness approached, Tanaka was tipped off about an American surface force in the vicinity consisting of two cruisers and four destroyers. If so, that threat could be handled by Vice Admiral Kondo who was returning to Lunga Point with *Kirishima* and two cruisers to conduct the mission that had been disrupted by Callaghan two nights earlier. Kondo's three gunships were screened by the light cruiser *Nagara* and six destroyers. In advance of the bombardment force, the light cruiser *Sendai* and three destroyers scouted for the reported Americans.[12]

At about 2317 that evening, *Atlanta*'s sailors near Lunga Point began hearing thunder in the distance, ignorant of the source. Unbeknownst to the Japanese, the two reported American cruisers were the battleships *South Dakota* and *Washington* under the command of Rear Admiral Lee, who was embarked on the latter.[13]

Lee's force had arrived by passing west of Guadalcanal and Savo Island, and then pivoted his column right around Savo to steer south into Ironbottom Bay. Lee then turned his six-ship column—led by the destroyers *Walke*, *Benham*, *Preston*, and *Gwin*, followed by *Washington* and *South Dakota*—in a westerly direction off the northwestern coast of Guadalcanal. At this point, his radar operators detected *Sendai* and the other ships of the Kondo's scout force.[14]

The light cruiser *Sendai* and destroyers *Shikinami* and *Unanami* evaded opening 16-inch salvos from the two American battleships. Besides changing courses, the Japanese expertly laid smoke to hide their movements. Perhaps the four destroyers in Lee's van might have better luck. With their FD fire-control radar and good night vision, they detected Kondo's screening force and prepared to attack. Meanwhile Kondo, intending for his bombardment group of a battleship and two cruisers to avoid the fray, steamed north of Savo Island with an objective of completing the mission of savaging Henderson Field.[15]

As the large rifles of *Washington* and *South Dakota* went silent, the 5-inch batteries on *Walke* and then *Benham* opened on the destroyer *Ayanami*. Five minutes later *Preston* attacked *Nagara* and the light cruiser's accompanying destroyers. *Gwin* and *Walke* did likewise. Once again, the Japanese demonstrated superior night-fighting skills as their gunfire and torpedoes landed punches on all four American tin cans. One Long Lance hit forward of *Walke*'s bridge, penetrating the magazine, setting off a violent explosion that lifted the ship up and snapped the bow off. Multiple shells hit *Preston*, penetrating the engineering spaces and this proved fatal. However, *Ayanami*'s counterbattery fire exposed that ship to *Washington*'s gunners who put steel on target. Before dawn *Ayanami* would join the growing fleet at the bottom of Ironbottom Sound.[16]

With *Walke* and *Preston* gone, Lee ordered his two remaining damaged destroyers to retire, and forged on with his two battleships. Of the two destroyers, only *Gwin* would survive as *Benham* could not overcome a torpedo hit near the bow.[17] Meanwhile, the Japanese scout ships that had engaged the Americans alerted Kondo to the presence of American battleships. Kondo, intent on completing his mission, altered his course to close on Lunga Point. Lee closed on Kondo.

Unfortunately for Lee, *South Dakota* suffered electrical distribution issues that evening, losing power during the initial engagement and again just before midnight. Once electricity was restored the second time, Kondo's force had come into visual range, a mere three miles off the battleship's starboard beam. Spotting the *South Dakota*, Japanese launched torpedoes and then the cruiser *Atago* confirmed the American warship's true identity with its searchlight. Other searchlights provided Japanese gunners with an inviting target. Luckily for *South Dakota*'s crew, several of the Long Lance torpedoes exploded early, fooling the Japanese to think they had landed fatal blows. However, shells of varying caliber did score hits within the battleship's forward mast structure, disabling radio antennae, gun directors, and radar housings. Minus sensor inputs, the battleship's 16-inch batteries fired only a few salvos. In contrast, the battleship's secondary starboard 5-inch 38 twin mounts gave a better account providing continuous counterbattery fire.[18]

From Point Lunga, Electrician's Mates McKinney, Oscar Ekberg, Vincent Burke, Warren Cantrell, and Warren McDorman had been recruited the previous afternoon to crew a large 60-inch searchlight installed on the beach to spot any nighttime Japanese waterborne infiltration. Unfortunately for the quintet, some foxhole digger had severed the cable leading from the diesel generator to the searchlight, so their posting was pointless that evening except that they were provided a viewing perch for the distant naval action. "We were fascinated spectators," McKinney recalled. "We could see ships firing at each other."

Beyond *South Dakota*, *Washington* stalked in the darkness. With *South Dakota* drawing Japanese searchlights and gunfire, Lee and *Washington*'s Commander Capt. Glenn Davis understood what they were up against. They assigned one 5-inch mount to fire star shells to illuminate Kondo's force and a second mount to target *Atago*. The remaining two 5-inch mounts and three 16-inch turrets were directed to pour their rounds into *Kirishima*. Subsequent analysis claims upwards of 20 16-inch shells and a similar number of 5-inch rounds struck the modernized battleship. Rounds striking the hull below the waterline had the same impact as torpedoes. Meanwhile *Washington*'s turret dedicated to *Atago* poured an estimated 40-plus 5-inch shells into Kondo's flagship. *Atago* withstood the barrage and maneuvered westward with other undamaged Japanese warships to reposition for a torpedo attack.[19]

Leaving his crippled battleship in his wake, Kondo's remaining two cruisers from the bombardment group swung to launch torpedoes at 0013 at an evasive *Washington*, and attempted to engage with his guns to little effect. Meanwhile, *South Dakota*'s damage control teams worked to extinguish some 23 major fires as the battleship cleared the area. Determined to avenge the loss of his capital ship, at 0025, Kondo summoned the combatants assigned to the sweeping and screening forces along with Tanaka's inbound transport convoy to rendezvous at a point six miles to the north of Cape Esperance. Two of Tanaka's destroyers joined what proved to be a fruitless effort to sink *Washington*. Dodging several Japanese torpedo attacks, Lee and Davis escaped to the south. Though Lee would have loved to have placed the remnants of Tanaka's remaining convoy in his gunsights, he correctly understood that the sacrificial efforts of his six-ship surface force had delayed the Japanese timetable and prevented the bombardment of Henderson Field. With that, the Japanese reinforcement mission was doomed.[20]

As the thunder and distant flashing ceased for those observers on the beach, the *Atlanta* electricians had no idea who had won. Later they would learn that *Kirishima* would pass beneath the waves before dawn about 10 miles west of Savo Island as three Japanese destroyers stood by to recover her survivors. To McKinney, the Japanese battleship loss represented some redemption for the destruction of *Arizona* and *Oklahoma*: "Thus the enemy lost two battleships over the weekend to cancel out the two we lost at Pearl Harbor on December 7."[21]

However, in the dark hours before dawn, all that could be heard was the engines of motor torpedo boats off the coastline. As the first light glimmered to the east, the *Atlanta* electrician saw in plain view the four remaining Japanese transports that had beached themselves in the early morning hours. Grounded, the four vessels were sitting ducks to the swarms of aircraft that had risen from an unscathed Henderson Field. Japanese soldiers embarked on these vessels jumped into the water hoping to either be picked up with small boats or swim ashore. Many would be killed by the concussion of bombs exploding on impact in the vicinity of the doomed ships. McKinney would later write the "word came down from the beach that Jap bodies in the water were too numerous to count."[22]

Marine Corps and Army artillery joined in. Meanwhile two Navy 5-inch 51 caliber guns of World War I vintage had been set up along the shore but lacked gun crews to fire them. Upon hearing the plea for volunteers to crew those naval artillery pieces, several of *Atlanta*'s senior petty officer gunners who had experience training on those weapons readily stepped forward.

Observing the Japanese transports some 20,000 yards down range hurriedly off-loading cargo and soldiers, *Atlanta*'s gunners went to work. Mustin recalled: "… they just opened fire on her and, of course, began hitting very quickly. Not much adjustment is required when the target is plainly in sight and is not moving and they just shot it to pieces—just shot it to pieces." Shells from the two guns ripped into the closest *Maru* until the vessel's stern slipped under leaving her bow exposed at a very steep angle.[23]

Further contributing to the Japanese carnage, the destroyer *Meade* appeared. A recently commissioned *Benson*-class destroyer featuring four 5-inch 38 single mounts spent some 42 minutes raking the remaining three transports and showered the shoreline with 5-inch and 40mm shells. The four transports ablaze represented a symbolic conclusion to this most recent Japanese effort to recapture Guadalcanal. Some 2,000

The *Kinugawa Maru* was one of the Japanese cargo ships pummeled on November 15, 1942. (Archives Branch, Naval History and Heritage Command, Washington, DC; 80-G-K-1467)

Japanese troops made it ashore with nominal supplies and subsistence to reinforce their countrymen, a fraction of what was needed to drive the Americans off the island.[24]

However, the opposing Americans had little idea of the exact number of troops that had gotten ashore in the predawn hours of November 15. To reinforce the Marine defensive positions, refugee platoons of about 15 to 20 *Atlanta* sailors were formed under the leadership of the surviving junior officers and sent to positions around the perimeter. Concerned that a Japanese assault could break through and reach the field hospital, medical staff warned patients that they might have to defend themselves. With his broken right hand and arm immobilized with a temporary splint, Lt. Jim Shaw was asked if he had a pistol. Shaw responded: "Affirmative but my right hand is broken ... what the hell am I supposed to do with the pistol?" The inquirer curtly retorted: "You'll learn in a hellava hurry to use your left hand."[25]

To the relief of all, no attack materialized that night and on the next morning Shaw and numbers of other wounded sailors were placed on a cart to be hauled by a tractor to Henderson Field. Halfway through the trip the tractor stopped due to a Japanese artillery attack. Air evacuation assured *Atlanta* wounded superior medical care away from the fighting and malarial jungles of Guadalcanal. For the first leg of the journey, the wounded were shuttled to Espiritu Santo. Van Ostrand Perkins arrived with apprehension. Back on Guadalcanal he was confined to a stretcher with a tag detailing the nature of his wounds with a note "leg must be removed." However, since Perkins had lost so much blood, the doctors at Cactus decided to defer on the amputation and pass that determination on to the medical experts at Espiritu Santo. Perkins would keep his leg.[26]

Back on Guadalcanal, with the threat of a Japanese offensive diminishing, General Vandegrift proceeded to plan his own offensive operations to expand the American footprint. *Atlanta*'s sailors were released from defensive duties. The electrician's mates manning the searchlight battery, having repaired the cut cable, enjoyed the novelty of being armed with rifles. For target practice they took aim and fired at coconuts floating offshore. When not on duty, McKinney hunted for souvenirs, once collecting a helmet off a decaying enemy carcass. On one walk he came across a holding camp for Japanese POWs. Looking at about three dozen docile former soldiers of the Imperial Japanese Army, McKinney noted the heavily armed guard. Talking to a staff sergeant, McKinney learned that the guard was armed more to protect the prisoners from fellow Marines.[27]

Assigned to the Defense Battalion, the electrician's mates ate twice a day with the Marines. Mealtimes and locations were altered daily as a predictable gathering would make for an inviting target for Japanese artillery. McKinney enjoyed the pancakes and vegetable soup. In contrast, those who stayed at the *Atlanta* camp compound were offered the normal Navy routine of three servings a day. Those sailors ate with mess kits that the cruiser *Portland* provided while that damaged cruiser waited at Tulagi for an eventual tow to Australia. When not eating, those sailors not assigned to duty with the Marines dug latrines, stood two-hour guard shifts, and helped off-load barrels of gas and oil drums that were delivered daily from *Bobolink*. As with the Marines, the sailors were told to take Atabrine to combat the malarial bites of mosquitoes. McKinney took his prescribed dosage and more, and developed a heavy yellow skin sheen. Unfortunately, the rumor that the pills caused sterility persisted and shipmates who came down with the disease later confessed that they had not taken the pills.[28]

After a few days, the rumor mill spread word that the survivors off *Atlanta* and the other sunken destroyers would finally be evacuated. Anticipating *Libra* and the *Betelgeuse* to arrive on Friday November 20, the crew—minus a few who volunteered to stay at Cactus—gathered on crystal clear morning to embark on a small flotilla of small craft to charge out into the calm waters of Ironbottom Sound, hoping to spot the two cargo ships on the horizon. McKinney, lying on the bottom of a landing craft under the direct sun for several hours cursed the apparent screw-up and began to think he should have volunteered for duty ashore.[29]

The two ships did arrive the next day. Signalman Striker Harvey remembered embarking the *Betelgeuse* with Captain Jenkins with the rest going to the *Libra*. Part of the *Betelgeuse* contingent, McKinney recalled that he and his shipmates would have to earn their passage away from Guadalcanal by unloading the ship. First off came the drums of aviation gasoline that had been stored on deck. Once the five hatches were clear of that volatile cargo, they were opened to unveil the next thorny item to off-load—spools of barbed wire. *Atlanta*'s cargo handlers gingerly handled the wire that was hoisted out and lowered onto awaiting boats using the ship's cargo booms. The barbed wire heading ashore marked the end of the first day of off-loading, and the two ships got underway for an overnight in the less exposed waters of Tulagi. The next two days were spent mostly hustling 500-pound bombs into cargo nets for transfer over the side to awaiting small craft. Returning to Tulagi on that third night, *Betelgeuse* plowed over a sandbar. If the ship failed to back off the underwater obstruction overnight, the morning's light would expose the ship to Japanese observers in the hills overlooking Henderson Field. Air or an undersea attack could be anticipated. *Atlanta*'s sailors, appreciating the potency of enemy ordnance—especially on a ship loaded with ammunition—needed no motivation to get back to work to lighten the load. McKinney later wrote:

> All hands worked four hours on and four hours off. This kept enough men at work at all times to generate maximum effort. It was grueling effort and every man's heart was in his mouth. We all knew what one enemy torpedo could do. I recall that at the end of my first four hour shift I fell into a bunk fully dressed and wept in sheer exhaustion and despair.[30]

The *Atlanta* cargo handling shifts finally cleared the holds of the 500-pound bombs and reached the consignment of 3-inch mortar shells that were packaged in clusters of three. With the end of the off-loading in sight, the deliriously tired sailors sang and shouted as they flung the clusters "into the cargo nets as if they were so much firewood." With her cargo offloaded, the cargo ship successfully backed off the bar.[31]

Escorted by the destroyer *Mustin*, *Betelgeuse* left for Espiritu Santo on Wednesday November 25, with *Libra* departing a day later. Ironically, the destroyer named for Lloyd Mustin's father would be the only Mustin departure that day. The son remained, having received orders to serve as Captain Greenman's operations and intelligence officer.

On December 1, *Atlanta*'s survivors embarked on *Betelgeuse* and *Libra* arrived at Nouméa. En route they were treated to a full Thanksgiving dinner during the first leg of the trip to Espiritu Santo. Petty Officer

McKinney called *Betelgeuse* a "good feeder" noting that each table "sported a soup bowl with a pound of butter in it." After a two-day pit stop at Espiritu Santo, the two cargo ships continued on to Nouméa, escorted by *Mustin* and the damaged *Aaron Ward*.

Arriving at the southeastern New Caledonian port, *Atlanta*'s able-bodied survivors left the cargo ships and were sent ashore where they lined up to receive new sets of "dungarees, skivvies, shoes, blankets, razors, soap, etc." Trucks awaited and soon the sailors arrived at the "receiving ship." Whereas "receiving ship" usually meant just that—typically an obsolete vessel berthed at a port that served as a transfer point for transient sailors—in this case, the term applied to a tent city up on a nearby hill. Though under the charge of Lieutenant Mack, for the most part the sailors managed themselves, keeping the grounds tidy. Unfortunately, it was at this location where sailors who had passed on taking Atabrine at Guadalcanal came down with the symptoms of malaria. About a fifth of the present crew had to receive medical treatment. Some 40 bags of mail arrived. Before the letters from home could be distributed, a small detail sorted the correspondence into three piles: Dead—Living—Wounded.[32]

Rumors about what the Navy had lined up for the survivors ranged from shipment to Australia for rest and recreation, to being held at Nouméa as a replacement pool for sailors killed in action on other ships. On the first anniversary of the attack on Pearl Harbor, trucks retrieved *Atlanta*'s survivors and brought them down to the waterfront in preparation for the formal disbanding of *Atlanta*'s crew two days later, in anticipation of the arrival of the SS *President Monroe* which had been chartered by the Navy to serve as a transport. For many on Nouméa's waterfront, the news of *President Monroe*'s arrival meant a trip to San Francisco and then 30 days of survivors' leave. It would be a long trip, made longer due to the need to stop at New Zealand to pick up additional passengers—wounded sailors, soldiers, and Marines who had been treated at medical facilities around Auckland. Not all of the *Atlanta* veterans boarded the *President Monroe*. Those having critical skill sets were ordered to remain. In one case, an *Atlanta* sailor of French-Canadian descent who spoke the language fluently was retained and instructed to live out in town in civilian attire to continually measure local sentiments about the Americans who had established a major base at this Vichy French colony.[33]

For those who earned the trip home, the next three weeks would be spent on a relatively new American President Lines passenger and cargo ship that had been chartered by the Navy to provide such service. Arriving at Auckland, the passengers were awed by the beauty of the mountains rising off the ocean during the inbound passage. For the *Atlanta* veterans, Auckland represented the first major city port call since Honolulu five months earlier and they were delighted to learn upon arriving at the Princess Wharf that for the next four days they had "open gangway" privileges—they could come and go as they pleased![34]

Petty Officer McKinney remembered going out into the city with a few of his shipmates and finding a small luncheonette-type establishment. He heartily consumed an order of soup, steak, and eggs, followed up with a large slice of pie topped with ice cream, all chased down with four glasses of milk. The small contingent of *Atlanta* veterans then flagged down a city bus that had the U.S. naval base on its route. The base featured a hospital camp where many of the light cruiser's seriously wounded had been flown to. Among the recovering, they may have seen Seaman 1/c Franklyn Reed who had several pieces of shrapnel removed. Naval Academy classmates Lieutenants Perkins and Shaw had been reunited as their respective limb injuries would take several months to heal. For *Atlanta*'s wounded, the influx of visitors was a morale boast.[35]

At the naval base, McKinney and his cohorts were also able to receive some backpay. With some spending money in hand, the sailors went on a shopping spree, finding the locals very welcoming and appreciative of the American servicemen. To McKinney's amazement, he discovered that the young New Zealander ladies were more than welcoming. With many young New Zealand men serving with Commonwealth forces in the Middle East and on other fronts the electrician's mate found "… they were quite frank and got to the point quickly." With such hospitality, a number of the sailors found no need to return to the ship during their 96-hour stay.[36]

As *President Monroe* prepared to depart, caravans of vehicles arrived from the naval base, carrying the wounded who were deemed stable enough to make the two-week voyage. McKinney recalled that the number that they were supposed to transport back had been about 500 but many more boarded and space had to be found in some of the holds to berth some of the 300 unanticipated extra passengers. Some 75 bed patients were placed in more civilized staterooms. The able-bodied *Atlanta* veterans were each assigned a bedridden running mate. In McKinney's case, he was paired with a machinist's mate off another cruiser who had been crushed by falling machinery and had been covered with a full body cast. McKinney tended to the injured sailor's needs, going down to the mess line to draw a ration of food and feeding him, then cleaning the used utensils for his own use.[37]

McKinney's best memory of the cruise was Christmas where the ship's crew hosted a party. McKinney recalled there were two Marines, each with an amputated leg, who played the piano together to provide holiday tunes. Presents were distributed—wrapped packages containing shaving soap, toothpaste, and candy. "Nothing much, but our hats were off to the people who remembered that everybody likes to have a package to unwrap at Christmas." A week later, they celebrated New Year's Day by arriving off the Golden Gate. Departing the ship, the able-bodied *Atlanta* veterans were sent to the receiving station at Treasure Island, a landfill in the middle of the bay that had been the site of the 1939 Golden Gate International Exposition. There they were assigned a bunk and a 72-hour pass. Upon return, they would receive papers authorizing their 30-day survivor leave to visit family, and orders for their next assignment. Many of the sailors took on the added chore of visiting the next-of-kin to those who were killed or were wounded to extend condolences and comfort. In the case of Ed Corboy, upon his return to his hometown he visited with the editorial staff at his former employer, the *Chicago Tribune*. While on leave, Corboy would draft a series of articles under the title of "Saga of Mighty *Atlanta* Told By Chicagoan" that honored the service of his fellow shipmates.[38]

At the time of their arrival, the American public had yet to learn of the loss of the *Atlanta*. For operational security purposes, the Navy typically did not immediately announce the loss of ships so as not to announce to the enemy what chess pieces no longer remained in play. However, the families of *Atlanta*'s crewmembers became immediately aware that the ship had sustained heavy damage during the great November victory that was reported in papers across the country as the reporter, who had been embarked on *Atlanta*, discussed that an American ship had been hit, suffering numerous casualties noting, "Even the ship's mascot. The little dog Lucky, died at his post—not a scratch on him, probably concussion killed him."[39]

Reading the article in the *New York Herald Tribune*, Elizabeth "Betts" Perkins felt sick, wondering of the fate of her husband and his shipmates. She later reflected:

> For the first time I was struck with horror at the brutality of war. To build a thing of beauty, to fill her with the finest young men the Navy could muster, and to send them out to be killed or maimed! How fruitless, how unconscionable—the ultimate stupidity of mankind, the ultimate horror of womankind. I came to hate, unconditionally, for the first time in my life. The Japanese were the focus of my hate.[40]

Over the coming days, Mrs. Perkins scoured the *New York Herald Tribune* for further details and on November 28, an article appeared with a headline that read "Detailed Story of U.S. Victory Over Jap Fleet in Solomons" with a subtitle stating, "Naval Officers Describe 3 Day Battle of Sea and Air Forces." In the article, a quote caught her attention: "It was like a scene in the movies when somebody turns out the lights in a barroom and everyone starts shooting at once." The quote, attributed to "Lieut. VanO Perkins of Greenwich, Conn" caused the young bride to scream up to her parents, "He's alive, he's alive. Oh thank God he's all right!"

Betts Perkins's exuberance was suddenly tempered by the realization that others could expect to hear devastating news. Grace Smith, the wife of the communications officer called: "Phil has been killed!" Perkins drove over to console the bereaved young women only to arrive to see her talking on the phone with Lieutenant Pierce's mother. Smith exclaimed: "Oh God, Jack has been killed." Grace Smith, setting aside her own grief, hopped in the car with Betts Perkins to be with the deceased officer's mom. Whereas no support network existed prior to the engagement off Guadalcanal on the morning of November 13, one quickly formed in early December as family members began to learn the fates of their loved ones. Commander Emory's post-battle action report noted that, of the 735 crewmembers, 158 had been killed and 14 enlisted were still missing—approximately 23 percent of the crew. Within a wardroom of 45 where 19 officers had made the supreme sacrifice, the percentage killed topped 42 percent. Heavily censored letters began to trickle back. In some cases the correspondence caused further angst as censors cut out mentions of wounds. Adding to the pain of the news of the loss of officers and enlisted sailors on the *Atlanta*, was news of far more extensive loss of life on sister ship *Juneau*. In the days leading up to the celebration of the birth of Christ, gatherings were occurring at places of worship across the land to remember those who their lives in combat. Remarkably for the Atlanta region that provided a fair number of the light cruiser's recruits, the number of home-losses were slight. By mid-January the local papers had only identified two fatal casualties from the area—Ensign Ashley Davis Morris and Chief Warrant Officer O. O. Cleveland.[41]

CHAPTER 14

Atlanta Reborn

Georgia's capital city took great civic pride in the light cruiser that represented the city in combat and for good reason. During the nearly 11 months that *Atlanta* had been in commission, the ship had been awarded five battle stars and a Presidential Unit Citation for her action on the morning of November 13. Upon learning of her loss, city leaders looked to Houston which had lost a cruiser named for that Texas city earlier in the war and were impressed how the local citizenry backed a war bond drive to raise funds to underwrite a replacement hull.

Following a meeting in early January with members of the Atlanta Chamber of Commerce; the state war bond staff; and representatives of Atlanta's two major newspapers, John L. Conner was nominated to chair the fundraising effort. As the president of the Southern Federal Savings and Loan Association with strong institutional connections, Conner kicked off the drive by purchasing $10,000 of bonds, and wasted no time in reaching out to the White House to gain a commitment that a new cruiser would be named for the city. The Secretary of the Treasury Henry Morganthau immediately wired his delight in hearing of the bond drive effort. With regards to naming the new ship, that matter had to await the return to the nation's capital of the Secretary of the Navy, Frank Knox, who had oversight over the ship naming process. However, given that Georgia Congressman Carl Vinson, chairman of the House Naval Affairs Committee, had already speculated that the name would be given to one of the five 10,000-tonner cruisers currently under construction, a new *Atlanta* was a foregone conclusion.[1]

Confirmation of the naming came in early February as Congressman Vinson notified the Atlanta's newspapers that the hull known as CL-104 would become the new *Atlanta*. To spur public interest in the bond drive, the *Atlanta Journal* contacted the White House to obtain pictures of the ship under construction. Unfortunately, not much would be forthcoming as the ship had only been laid down on January 25. Like the previous *Atlanta*, the new ship would be a product of the Garden State, as the New York Shipbuilding Corporation yard at Camden received the construction contract. Unlike CL-51 which was the first of her class, CL-104 would be the next to last of 27 *Cleveland*-class light cruisers. Displacing over 14,000 tons, the new warship would host a main battery of 12 6-inch 47 caliber guns packed into four triple turrets with two forward and two aft. A secondary battery of six 5-inch 38 twin mounts added to the ship's firepower as did the numerous 20mm and 40mm mounts.[2]

Pictures aside, the campaign to sell $35,750,000 in war bonds began to steamroll in February 1943, following the return from leave of several of the native-born veterans of the lost *Atlanta*. Throughout the month, an average of a million dollars in war bonds for the building of the new *Atlanta* was raised per day. Whereas the initial deadline to finish the campaign had been set for April 15, the pace of bond sales enabled the fundraising committee to conclude the effort a month early. On March 3, a headline in the *Atlanta Constitution* proudly boasted "Cruiser Campaign Sweeps over Top."[3] This allowed Conner to move forward with a planned War Bond Victory Celebration event at the municipal auditorium on March 12. On that evening, over 8,000 individuals crammed into the auditorium that day to hear an appreciative speech from Secretary Knox. Conner then read a telegram he received from Washington:

> My message to the people of Georgia and of Atlanta is "well done." You and your faithful co-workers have given the most effective answer to our adversaries. Hearty congratulations.
> Franklin D. Roosevelt

Conner later boasted that the message "received the largest applause in the in the history of the Atlanta auditorium." Conner then presented Knox with a symbolic check of $63,000,000, representing the amount of war bonds Georgians had purchased. For the presentation, Conner and

Standing second from the left, Secretary of the Navy Frank Knox receives a $63,000.000 check from John L. Conner, Chairman of the City of Atlanta, Georgia, from the War Bond selling campaign to pay for a new ship, USS *Atlanta* (CL-104). Looking on are Rear Adm. George D. Murray (left), and Capt. Samuel P. Jenkins (right), who commanded Atlanta (CL-51). (Archives Branch, Naval History and Heritage Command, Washington, DC; NH 70035)

Knox were bracketed by CL-51's former skipper Capt. Samuel Jenkins and Rear Adm. George Murray.[4]

Once again, Mrs. John R. March—Margaret Mitchell—served as the ship's sponsor and on February 6, 1944, she once again cracked a bottle of Champagne against the bow of a warship under construction and a new *Atlanta* slid down the ways into the Delaware River. Mitchell would return to the Delaware Valley for the commissioning on December 3, 1944. Capt. Bayard H. Colyear accepted command on behalf of a crew of over 1,200. Among those standing at attention that cold day were his new medical and dental officers. Cdr. Carl C.

Garver would serve on the new *Atlanta* for the next 12 months.[5] Lt. Cdr. Robert Erdman would tend to the crew's dental needs until March 1946.[6] Five other former *Mighty A* sailors stood at attention among the ranks of the crew after having their requests to be plank owners of the new cruiser granted.[7] As with the previous *Atlanta*, the newly commissioned cruiser spent time on the East Coast to test out equipment and train the new crew before heading to the Pacific and reaching Pearl Harbor on April 18, 1945. She arrived at Ulithi Atoll on May 12 to join Task Force 58. For the remaining three months of the war, the light cruiser provided anti-air defense for the fast carriers that were launching airstrikes against the Japanese home islands. *Atlanta* also closed on the Japanese islands of Honshu and Hokkaido to engage shore targets with her 6-inch batteries.

When hostilities ceased, *Atlanta* joined the American armada in Tokyo Bay for two weeks at the end of September before returning to Seattle. During the postwar period, *Atlanta* conducted two more Far East deployments. Following a reservist training cruise in early 1949, the light cruiser was decommissioned. Unlike many combatants that had been laid up following World War II, *Atlanta* would not be reactivated for the Korean War. While four of her *Cleveland*-class sister ships—*Galveston, Little Rock, Providence*, and *Oklahoma City* would be modernized into guided missile cruisers to combat the threat from Soviet naval aviation, *Atlanta* would be struck from the Navy register on October 1, 1962. However, instead of being scrapped, the relatively young hull was re-envisioned as a platform to carry experimental superstructures to be subjected to explosive testing. Thus, *Atlanta* had a second career during the mid-1960s as a uniquely configured target ship. Struck again from the Navy list on April 1, 1970, the former light cruiser met a fate similar to her predecessor in that she was sunk. However, there were no casualties as a result of the explosive test conducted off San Clemente Island on October 1, 1970.

In the spring of 1979, the Metro Atlanta Council of the Navy League of the United States and its "USS *Atlanta*" committee led by former mayor Ivan Allen Jr. and the Atlanta Chamber of Commerce Military Task Force, sent a petition requesting a new USS *Atlanta*—signed by 301 business and civic leaders—to Secretary of the Navy W. Graham

As with the first and second *Atlanta*, the fourth and fifth ships to bear the name had radically different appearances. (Archives Branch, Naval History and Heritage Command, Washington, DC; NH 98879; DN-SN-82-05062)

Claytor, Jr. It probably didn't hurt the effort that the sitting president hailed from the Peach State and had trained to serve in the nuclear navy.

In the 1970s, the Navy named its new attack submarines for cities and on August 17, 1978, the keel of a *Los Angeles*-class submarine SSN-712 was laid at Newport News Shipbuilding. On June 11, 1979, Claytor sent Mayor Emeritus Allen a letter notifying him that SSN-712 would be named *Atlanta*.[8] Sadly, the sponsor of the third and fourth *Atlanta* would not be available to perform such a role for the fifth. Margaret Mitchell had tragically passed from the scene three decades earlier on August 16, 1949, after she was struck by a drunk driver as she crossed Peachtree Street in Atlanta to go see a movie. In her place, Colleen Nunn, the wife of one Georgia Senator Sam Nunn, had the honor of launching the new submarine on August 18, 1980, and participating in her commissioning on March 6, 1982. Cdr. Robin J. White served as the new *Atlanta*'s first commanding officer. The nuclear-powered "attack boat" designed to tangle with the Soviet undersea fleet as fictionally illustrated by Tom Clancy's *The Hunt for Red October*, would ably serve with the

Atlantic Fleet for nearly two decades, except for a grounding incident in 1986 during passage through the Strait of Gibraltar that damaged her sonar and punctured a forward ballast tank. The fifth *Atlanta* would be decommissioned on December 16, 1999.[9]

VIPs attending the christening and commissioning ceremonies included veterans from the third USS *Atlanta* of Guadalcanal fame. While some were discharged from the service due to injuries sustained on the early morning of November 13, 1942, others continued to fight with other naval units as the war continued. Not all survived. Van Ostrand Perkins, having received orders to serve an assistant damage control officer on the cruiser *Birmingham*, found himself off Leyte with Task Group 38.3 on the morning of October 24, 1944, when a Japanese dive-bomber scored a direct hit on the nearby light carrier *Princeton*. Though structural damage was light, the exploding bomb set off a chain reaction of burning gasoline that eventually would prove fatal. To contain the flames, nearby escorting destroyers and *Birmingham* took turns alongside as fireboats. As *Birmingham* come alongside again to douse the burning carrier, the flames reached *Princeton*'s magazines. The resultant blast killed Perkins and 232 of his shipmates and wounded 426 others.[10] In a twist of fate, Van Ostrand's widow Betts would remarry after the war to Van Ostrand's best friend and *Atlanta* shipmate, Jim Shaw. Shaw had lost his wife earlier during the war from a deadly combination of strep throat, scarlet fever, and pneumonia.[11] Shaw had his own close call with the grim reaper as he had been assigned to the carrier *Bunker Hill* off Okinawa on May 11, 1945, when that ship was hit by two Kamikaze aircraft, killing at least 346 sailors. Shaw would earn a Bronze Star that day as his antiaircraft guns fended off additional attacks as damage control crews contained the flames.

Shaw would be among those from *Atlanta*'s wardroom who would attain flag rank. However, his career path would be unconventional. In the immediate postwar years, he helped to chronicle World War II for Samuel E. Morison. He did command the destroyer *Waldron* which had been brought out of mothballs following the outbreak of war in Korea. During his two years in command, *Waldron* served with the Atlantic Fleet. He then drew more unconventional tours as a speechwriter for the Chief of Naval Operations and Secretary of the Navy, technical adviser for the film *The Caine Mutiny*, and duty as naval attaché to The Hague. In retirement, he would work for the Humane Society of the United States.[12]

Shaw's assignment to the pre-commissioning crew of *Bunker Hill* as the assistant gunnery officer was likely not random. Two months prior to Shaw's arrival, Lieutenant Commander Nickelson reported aboard that carrier as the gunnery officer. Eventually, Nickelson received orders off *Bunker Hill* in April 1944, to serve as the executive officer on the pre-commissioning crew of the heavy cruiser *Chicago*. Shaw would fleet up to fill Nickelson's shoes as *Bunker Hill*'s gun boss.

Following his XO tour in *Chicago* which remained deployed in the Far East following the surrender of Japan, Nickelson served ashore in Washington and at the Naval War College. Following command of the attack cargo ship *Winston*, he served as professor of naval science at the University of Mississippi NROTC. He served as a deputy chief of staff for the Caribbean Command before retiring in the 1950s as a captain.[13]

Another *Atlanta* ensign would make his way to *Bunker Hill*. Gerald Colleran would serve as one of Nickelson's division officers until departing in May 1944 with orders to perform similar gunnery officer duties in the cruiser *Springfield*. Perhaps his exposure to naval aviation inspired Colleran to apply for flight school after the war. Earning his wings of gold in November 1946 placed him on a career path that led to eventual command of the carrier *Bon Homme Richard* for combat deployments during the Vietnam War. Following a tour ashore in the Pentagon, Colleran retired as a captain on May 1, 1970, due to a physical disability.[14]

Yet another from the *Atlanta* wardroom who went on to make the Navy a career via the *Bunker Hill* was Ens. William H. Mack. He reported to the carrier a month prior to Nickelson's departure and eventually filled Shaw's old billet as assistant gunnery officer. His career in the surface navy prospered in the 1950s with commands of a troopship, a destroyer, and a destroyer leader named for Vice Adm. Willis "Ching" Lee. Promoted to captain on July 1, 1958, Mack would serve on tours in the Pentagon and at the Naval War College before returning to sea to command the *Cambria* and then Destroyer Squadron Two.

Both Captain Jenkins and Commander Emory continued to serve through World War II and would both be allowed to retire as rear admirals, to be otherwise known as "Tombstone Admirals" as you were entitled to place that rank on your grave marker. Recalled to the nation's capital, Jenkins spent a year on an assortment of chores to support the war effort to include traveling to Atlanta to support the bond drive effort. In 1944, he rejoined the fight in the Pacific as the commodore of various transport squadrons responsible for landing troops and supplies at Leyte, Lingayen Gulf, and Okinawa. Following the war, he assumed responsibility for ships being placed into Fleet Reserve status in the Columbia River as he, and other Naval Academy alumni were subjected to investigation by the Office of Naval Intelligence for membership in a secret society called the "Green Bowl" that allegedly corrupted the Navy's assignment system to draw career-enhancing orders for its membership. He retired on October 1, 1947, the same date a report was signed out by Rear Frank J. Lowry, absolving the Green Bowlers of any maleficence.[15]

Whereas Jenkins returned home for shore duty, Emory received orders to command the oiler *Neches* for seven months in 1943. He then received orders for duty in Washington, DC, to the headquarters of the Naval Transportation Service. Subsequently, he received command of the attack transport *Grafton* which was commissioned on January 5, 1945. Now Captain Emory delivered over 1,000 Seabees to the Western Pacific and would endure enemy air attacks off Okinawa. In one of Emory's more interesting taskings, *Grafton* transported 1,000 Japanese POWs from Okinawa to Saipan. Following the Japanese surrender, *Grafton* brought back returning sailors, soldiers, and Marines in Operation *Magic Carpet*. Returning to his hometown of Seattle, Emory had back-to-back tours, first with the Thirteenth Naval District and finally as the professor of naval science at the University of Washington, before retiring.[16]

Once *Atlanta*'s pre-comm acting commanding officer, Lt. Cdr. Norman Sears served as the First Lieutenant and Damage Control Officer, and was wounded on the morning of November 13 from fragments of a bursting shell. Despite his serious wounds, he made a valiant effort to direct repair efforts. Following his medical evacuation and recovery, he drew assignment as an executive officer of a floating dry dock on the

West Coast. From there, he joined the amphibious forces which proved to be a path to flag rank. During the final year of the war he commanded an LST (landing ship, tank) flotilla that supported landings at Peleliu and Okinawa. Assigned ashore in San Diego in the postwar period, Sears was rushed to the Far East in July 1950, following the North Korean invasion of South Korea, and reported to his new boss, Rear Adm. James H. Doyle, to serve as his Chief of Staff with Amphibious Group One. He arrived just in time to command a flotilla of supply ships during mid-July reinforcement landings at Pohang. Two months later, having helped plan one of history's boldest amphibious assaults, Sears commanded the advance attack force that steamed up Flying Fish Channel to land Marines on Wolmi-do, a land mass overlooking the approaches to Inchon. Sears would continue to serve with Doyle and eventually earn his stars before retiring.[17]

Another officer who would eventually wear the uniform of rear admiral in retirement was Jack Wulff. Having stayed behind with Lloyd Mustin at Guadalcanal, he was stuck there until May 1943. As Mustin served as assistant gunnery officer in *San Diego*, Wulff received orders to the pre-comm crew of the cruiser *Miami* that was under construction at the recently reopened Cramp Shipyard in Philadelphia. Much to Wulff's delight, a few months later Lloyd Mustin appeared as the gun boss for the cruiser commissioned on December 28, 1943. Unlike Mustin, *Miami* would be Wulff's home for the duration of the war and beyond. Returning to the States in late 1945, the cruiser spent the next year and a half as a training ship for naval reservists. With the release of its wartime crew from active duty, Wulff eventually fleeted up to become the CO before receiving orders for a year of shore duty in Chicago before returning to sea in 1948 in command of the destroyer *McCaffrey*. After a shore tour in San Francisco, Wulff again returned to sea in command of the attack transport *Diphda* which earned a battle star for service during the Korean War. After attending the Naval War College he would have one final command at sea and tour ashore before retiring in November 1959.[18]

As for Mustin, he left Guadalcanal in December to assume his old job on *San Diego* before receiving orders to *Miami*. After the battle of Leyte Gulf, Mustin received orders to serve on the staff of commander,

Battleship Squadron Two to work for Vice Adm. "Ching" Lee. When Lee received orders to return to *South Dakota*, Mustin followed as his flag aide. With the relationships he had fostered with senior Navy leadership and his mastery of gunnery, Mustin would have a most successful career in the surface navy, retiring as a vice admiral following a sea tour as commander amphibious force, U.S. Atlantic Fleet, and a tour ashore as the director, Defense Atomic Support Agency—an organization that had been established to monitor the safety of the nation's numerous nuclear weapon systems. Mustin retired on August 1, 1971, after serving in the Navy for 43 years.[19] Lloyd Mustin's interest in putting bullets on target continued. He served two terms as president of the National Rifle Association in the late 1970s and would serve with the U.S. Olympic Committee for shooting sports.[20]

A twisted fate awaited Girard "Pat" McEntee once he recovered from his hand injuries. On February 6, 1943, he reported to the cruiser *San Francisco* to assume duties as the gunnery officer—assuming responsibility over the gun crews that inadvertently fired salvos into his previous command. Under McEntee's supervision, *San Francisco*'s 8-inch batteries during the Marianas campaign leveled Japanese defenses on Saipan and Tinian. Leaving *San Francisco* in the autumn of 1944, now Commander McEntee returned for duty at the Naval Academy and after two years, retired by reason of physical disability.[21]

Lt. (jg) Paul Luddy was one of the *Atlanta* survivors to remain behind at Guadalcanal and take on the duties as first lieutenant at the PT base across the Sound on Tulagi. Eventually, he served as the Assistant First Lieutenant on the cruiser *Reno*. Following World War II, he left active duty to pursue a law degree at Boston University, and affiliated with the Naval Reserve. Once he passed the bar, he attended Navy Judge Advocate General School and returned to active duty as a legal officer until he retired in 1966 as a commander to serve as a probate attorney in San Diego.[22]

Of note, most the Naval Academy graduates of the *Atlanta* wardroom who if physically able, opted to remain in the service after the firing ceased. David P. Hall was one exception. After recovering from his injuries, he, along with a number of *Atlanta* survivors, would spend the duration of the war assigned to the light cruiser *Oakland*. Built on the West Coast, this first of an additional four *Atlanta*-type light cruisers differed in that port and starboard 5-inch 38 twin mounts were removed from the design to enhance stability. As the assistant gunnery officer, Hall remained with that ship during the campaign for the Gilberts, the Philippines, Iwo Jima, and Okinawa, eventually winding up in Tokyo Bay on September 2, 1945, with the American naval armada that was on hand to witness the surrender of the Japanese Empire. Following the end of the war, Hall transitioned to civilian life, remaining in the naval reserves for the next nine years.[23]

Two of the V-7 program graduates, Edward Corboy and Robert D. Graff, left the service after the war to work in media. Corboy, who had worked for the *Chicago Tribune* prior to the war, returned to work for Col. Robert McCormick, after publishing a serial on *Atlanta* with the *Trib* in 1943. One of Corboy's postwar accomplishments was working with McCormick in a lobbying effort to name the new airport on the outskirts of the city for Navy ace Edward "Butch" O'Hare. Cowboy worked to sell advertising for McCormick's paper, retiring in 1984 as the head of the *Tribune's* corporate advertising division. In contrast, Graff entertained the new medium of television and had a successful career as a television producer for the upstart American Broadcasting Company.[24]

Many of the enlisted sailors opted to remain in the Navy—some broke into the officer ranks. Boatswain's Mate Robert Smith retired in 1957 at the rank of Chief Warrant Officer. Fireman Charles Dodd eventually earned a spot at the NROTC unit at the University of Southern California and earned a line officer commission. He would continue with a career in the surface navy retiring on June 30, 1958. A day later, Pharmacist's Mate Gerald Ashlock who earned his commission after the war, retired as a lieutenant commander. Pharmacist's Mate Fredrick Stauffer would continue looking after shipmates and Marines for the next 24 years, retiring as a Chief Warrant Officer.[25]

Boatswain's Mate Leighton Spadone, after spending four months in Auckland recovering from his injuries, returned to the United States and earned a commission. Now Ensign Spadone received orders to command an assortment of medium-size amphibious ships through the duration of the war. For his final tour of active duty before retiring as a lieutenant commander in 1959, Spadone commanded the high-speed transport *Cavallaro*. Spadone settled in Colorado Springs to serve as a sales manager at various auto dealerships. Electrician's Mate William McKinney, who also earned his ensign butter bars, joined the contingent of former *Atlanta* sailors assigned to *Oakland*. Following assignment to the battlecruiser *Alaska*, McKinney retired in 1952 as a lieutenant commander. Moving to Ipswich, Massachusetts, McKinney would have a productive career in the dairy business.[26]

As with the majority of those who served during World War II, most of *Atlanta*'s sailors returned to civilian life after the war as part of "the Greatest Generation." For example, after recovering from his injuries, Franklyn LeRoy Reed would return to Connecticut with Breta Walker—the woman he met while performing chauffer duty in New Zealand—and would have a happy marriage and a long career working for General Dynamics Electric Boat, building submarines for the Navy. Clifford Dunaway, who stayed behind at Guadalcanal to serve in PT boats for a year, returned home for well-deserved leave and then continued service on those small hulls in the Mediterranean until after the end of the war. Returning to Georgia, he married Grace Rayburn and gained long-term employment with Delta Airlines.

Decades later Grace Rayburn Dunaway served as the Matron of Honor during the launching of the fifth ship to be named for the capital of Georgia and in honor of the previous ships that had the name.

In 1992, the U.S. Navy and the National Geographic Society partnered to support a Robert Ballard exploration of Ironbottom Sound. The subsequent National Geographic documentary *Lost Fleet of Guadalcanal*, featuring introductory remarks from President George H. W. Bush, incorporated veterans from the opposing sides of the six-month struggle for the island and adjacent waters to humanize the chronological narrative as well as assist Ballard as his remotely operated underwater vehicles came upon the remnants of the various sunken warships. To provide commentary for the first phase of what has become known as the Naval Battle of Guadalcanal, Ballard welcomed Michiharu Shinya, a former Japanese sailor who had served in *Akatsuki* and H. Stewart Morehead, who as Rear Admiral Scott's operations officer, had recently embarked on the light cruiser and had been on the bridge that night. Ballard's team proved successful in locating the American cruiser and the punishment sustained that early morning of November 13, 1942, was gruesomely evident. Ballard's team also may have come upon Shinya's former ship *Akatsuki*. Unfortunately, since a sister ship *Ayanami* was sunk in the same vicinity during the second phase of the Naval Battle of Guadalcanal, Ballard hesitated to ascertain the exact identity of this particular Japanese warship. In the aftermath of the Ballard expedition, there have been follow-on surveys conducted on the wreck.[27]

In November 1998, a member of *Atlanta*'s wardroom made one final trip to Guadalcanal to pay homage to his fallen shipmates on the 56th anniversary of the loss of the light cruiser. Bob Graff traveled to the distant South Pacific island thanks to his stepson Christopher who presented the trip as a gift the prior Christmas. So for a week, Bob toured the island with Christopher who brought his son Kenny along. Upon coming to the numerous memorials scattered across the once war-torn landscape, the trio placed flowers and reflected, wishing everyone peace and harmony. On the next to last day of their trip, they chartered a diving boat skippered by a captain who knew the exact location where *Atlanta* rested and arranged for a floral raft to be towed to the site as well as the local Episcopal priest to join the party. Arriving at the site, Graff asked the boat skipper to turn off the engine and air conditioning so they could recite scripture. After the attendees took turns reading the biblical passages, Bob Graff pulled out a sheet of paper and shared some concluding thoughts:

> Dear shipmates and friends, those of us on deck today have come across half the world to be nearer to you in this anointed spot. You have called us and we have come to commune together more closely and to share our love. Others you left behind are less fortunate, they could not be with us. So we are only a handful able to come here, a tiny handful of those of us who are still afloat.
>
> On one horizon today sharp, verdant peaks of Guadalcanal pierce the sky. From the waters surrounding us, millions of javelins, reflected rays of the sun, blind us with your memory and pierce our hearts. Once a while back, we together, all of us made *Atlanta* what she was, pride of the fleet, the *"Mighty A"*. We were the youthful hope of the nation and the promise of mankind. Taking the world as

we found it, in our way and in our time, we tried to remake the world—more hope, more possibility and much larger community for happiness. That is what, years ago, brought us to Guadalcanal.

You did not fail our promise, nor did those of us here today fail you. In *Atlanta*, all of us went out so briefly beyond the bounds of our skin. We fused our passions then just as we hug you today, memory and myth. These few words floating in midair, this fragile plank with our call, these fruits and flowers and foliage are but mirages, passing impressions, of the past, frail appearances stronger than shattered steel. Bobbing wisps bind us together forever.

On the Japanese memorial to their dead on a sunny Guadalcanal hill, written in bronze on marble, we found this text: At this place repose all the spirits of those who sacrificed their lives in World War II in Guadalcanal and the entire Pacific area. This monument represents a requiem for their souls, and, I add, as does this brief ceremony at sea.[28]

Closing Note: On October 23, 2024, the Secretary of the Navy, Carlos Del Toro, announced that the future *Virginia*-class nuclear-powered attack submarine SSN-813 will be named USS *Atlanta*. The next *Atlanta*'s ship sponsor will be former Atlanta Mayor Keisha Lance Bottoms.

Endnotes

Chapter 1

1 Correspondence between Captain Bidwell and Mayor LeCraw from May 2 through to June, 1941; Admiral C. W. Nimitz memorandum to Bureau of Ships, June 8, 1941, *Atlanta* Pre-commissioning folder, Ships History Branch, NHHC.
2 "Launching Report, September 6, 1941, from Building Way #8 of Federal Shipbuilding and Dry Dock Company," *Atlanta* Pre–commissioning folder, Ships History Branch, NHHC.
3 "Navy Launches Two Cruisers," *New York Times*, September 7, 1941, 1, 42.
4 Raimondo Luraghi, *A History of the Confederate Navy* (Annapolis, MD: Naval Institute Press, 1996), 200–02, 210–15; J. Thomas Scharf, *History of the Confederate States Navy* (New York: Gramercy Books, 1996), 639–44; "History of USS *Atlanta* (CL-51) prepared by Office of Naval Records and History, Ships' Histories Section, Navy Department, USS *Atlanta* folder, Ships History Branch, NHHC.
5 Capt. Ralph Styles, USN (Ret.) interview by Vice Adm. Earl Fowler, USN (Ret.), Naval Historical Foundation Oral History Program, October 2001.
6 Charles V. Reynolds, Jr., *America and a Two Ocean Navy, 1933–1941* (PhD diss., Boston College, 1978), 214–21.
7 Navy Department Press Release of 15 February 1939, *Atlanta* Pre-commissioning folder, Ships History Branch, NHHC; Reynolds, 221–22.
8 Norman Friedman, *U.S. Cruisers: An Illustrated Design History* (Annapolis, MD: Naval Institute Press, 1984), 217–18.
9 Ibid., 222–23.
10 To clarify Navy gun nomenclature, 6" 47 means the gun barrel is six inches in diameter and the length of the barrel is six inches × 47, equaling 282 inches or 23 and a half feet long.
11 Ibid., 232–33.
12 Ibid., 236.
13 James D. Hornfischer, *Neptune's Inferno: The U.S. Navy at Guadalcanal* (New York: Bantam Books, 2011), 17; Friedman, 236.

14 Navy Department Press Release of February 15, 1939, *Atlanta* Pre-commissioning folder, Ships History Branch, NHHC; Fact sheet, *Atlanta* folder, Ships History Branch, NHHC.
15 "Shipyard Union calls 16,000 out on Defense Jobs," *New York Times*, August 7, 1941, 1, 12; Reynolds, Appendix.
16 Ship's Data Card, *Atlanta* file, Ships History Branch, NHHC.
17 C. L. Sulzberger, *World War II* (Boston, MA: Houghton Mifflin Company, 1966), 32–38.
18 Ibid., 282–83.
19 "Shipyard Union calls 16,000 out on Defense Jobs," *New York Times*, August 7, 1941, 1, 12.
20 "Shipyard Seizure by U.S. suggested at Kearny Strike," *New York Times*, August 8, 1941, 1.
21 Vice Adm. Harold G. Bowen, USN (Ret.), *Ships, Machinery, and Mossbacks: The Autobiography of a Naval Engineer* (Princeton: Princeton University Press, 1954), 208–09.
22 Frank L. Kluckhorn "U.S. Takes Over Kearny Shipyard, Open Tomorrow," *New York Times*, August 24, 1941, 1, 17.
23 Vice Adm. Lloyd M. Mustin, USN (Ret.) interview by John Mason, U.S. Naval Institute.
24 "Pioneer in Naval Aeronautics Dead," *The Evening Star* (Washington), August 24, 1923, 2.
25 1925 *Lucky Bag*; Captain Norman Scott Sears biography, Navy Department.
26 Emory Files, Operational Archives, NHHC.
27 1927 *Lucky Bag*, 250, 342; Loeser and Nickelson Files. Operational Archives, NHHC.
28 Elizabeth R. P. Shaw, *Beside Me Still: A Memoir of Love and Loss in World War II* (Annapolis, MD: Naval Institute Press, 2002) 70–73.
29 David F. Winkler, *Ready Then, Ready Now, Ready Always: More Than a Century of Service by Citizen Sailors* (Washington, DC: Navy Reserve Centennial Committee, 2014), 70.
30 Edward D. Corboy "The Log of the '*Mighty A*'" as told to Charles Leavelle, *The Evening Bulletin* (Philadelphia) March 22, 1943. The 12-part series first appeared in the *Chicago Tribune* and was syndicated nationally. Corboy hides his authorship. Author identified him through the process of elimination and a phone interview with him on November 19, 2002.
31 "Craighill biography" Z-files, Operational Archives, NHHC.
32 1921 *Lucky Bag*; Rear Admiral Campbell Dallas Emory Biography, Navy Department Library; Robert D. Graff oral history with author, dated December 27, 2003, 10.
33 Corboy, "The Log of '*Mighty A*,'" part 3.
34 Graff oral history, 6. In his oral history, Graff stated he reported to Captain Jenkins, but Jenkins did not arrive until early December.
35 Garner Survey from Robert Graff personal papers.

36 Clifford Dunaway, Sr., oral history, *Atlanta* History Center Veterans History Project Oral History, YouTube posted January 27, 2016. Dunaway would be the last surviving recruit from Georgia, passing away on November 5, 2017. Obituary *Atlanta Journal Constitution*, November 8, 2017.
37 *Chock "A" Block*, Vol. I, no. 1, August 6, 1942, 8.
38 William B. McKinney, *Join the Navy and See The World* (Los Angeles, CA: Military Literary Guild, 1990), 15.
39 Ibid., 12–19.
40 Ingersoll letter to Bowen of October 25, 1941, Box 2, Folder USS *Atlanta* CL-51 Launching and Trials, Entry 157, RG 80 General Records of the Department of the Navy, NARA, NY.
41 Memorandum, Box 2, Folder USS *Atlanta* CL-51 Launching and Trials, Entry 157, RG 80 General Records of the Department of the Navy, NARA, NY.
42 "Chronological Schedule of Principal and Governing Events to Effect Delivery by December 31st, 1941" dated September 17, 1941; C. A. Holderness memorandum to Rear Adm. H. G. Bowen, dated October 20, 1941, Box 2, USS *Atlanta* Folder, Entry 157, RG 80 General Records of the Department of the Navy, NARA, New York; W. C. Hemingway Memo to Bowen of December 3, 1941, Subj: Nov 41 Operations, Box 9, Harold G. Bowen Papers, Seeley G. Mudd Manuscript Library, Princeton University, Princeton, NJ.; Robert Graff noted in his oral history that the drydocking at Brooklyn occurred because the anti-fouling paint that had been applied at Federal Shipbuilding had failed. During a trial run the ship heeled over and someone observed "My God, the hull's rusting away!"

Chapter 2

1 1915 *Lucky Bag*, 105; The Green Bowl Society was exposed in a post-World War II investigation following accusations that the society promoted favoritism for its members as they advanced in their careers. See: "Lowry Report Absolves 'Green Bowl' of All Charges; 18 Admirals Listed." https://www.cia.gov/readingroom/docs/CIA-RDP58-00597A000100080005-0.pdf.
2 Jenkins Files, Operational Archives, NHHC; Rear Adm. Samuel P. Jenkins biography, Navy Department Library.
3 Shaw, 75–76; Graff oral history, 6.
4 Bowen, 166.
5 Flynn letter to Bowen of December 10, 1941, Box 9, Harold G. Bowen Papers, Seeley G. Mudd Manuscript Library, Princeton University, Princeton, NJ.
6 Ibid., 222. Unfortunately for Bowen, his dreams of going back to work full-time at Naval Research Laboratory never materialized as during the war he would seize seven more industrial facilities on behalf of the Navy to assure war production, Bowen, 203.
7 Shaw, 78–79.
8 McKinney, 19–20.

9 See NH 57450 USS *Atlanta* (CL-51) posted at www.history.navy.mil.
10 Graff oral history, 8; *Atlanta* Deck Log entry of December 24, 1941, NARA II.
11 "Navy Commissions New Type Cruiser," *New York Times*, December 25, 1941; Shaw, 80–81; *Atlanta* Deck Log entry of December 24, 1941, NARA II.
12 H. Charles Dahn, "The Death of the *Mighty A*," paper written December 28, 1958, Douglas St. Denis papers.
13 Graff oral history, 8.
14 McKinney, 21; Shaw, 79–80; *Atlanta* Deck Log entry of December 25, 1941, NARA II. When provisions arrived, one of the officers, typically a medical officer, would inspect the food for spoilage.
15 Clifford Dunaway, Sr. oral history, *Atlanta* History Center Veterans History Project Oral History, YouTube posted January 27, 2016; USS *Atlanta* Alumni Assn. Survey—C. B. Dunaway.
16 Franklyn LeRoy Reed oral history with Michael Willie of December 20, 2004, Library of Congress Veterans History Project.
17 USS *Atlanta* Alumni Assn. Survey—Leighton Spadone.
18 *Atlanta* Deck Log entry of December 29, 1941, NARA II; Maw C. Peterson, "Naval Courts-Martial," *Indiana Law Journal,* Vol. 20, Issue 2 (Winter 1945), 168–69. For very serious criminal activity, Captain Jenkins could refer a case to his immediate commander to convene a General Court-Martial of not less than five or more than 13 officers as members.
19 *Atlanta* Deck Log entries of January 6 and 29, 1942, NARA II.
20 Ed Corboy, "The Log of the '*Mighty A*'". (USS *Atlanta* Reunion Scrapbook, 1985) 5.
21 *Atlanta* Deck Log entries of January 2, 9, and 20, 1942, NARA II.
22 McKinney, 20, Corboy, 5.
23 *Atlanta* Deck Log entry of February 8, 1942, NARA II. 1MC stands for 1 Main Circuit.
24 Ibid.
25 Graff oral history, 14–15.
26 *Atlanta* Deck Log entry of February 10, 1942, NARA II.
27 Graff oral history, 12–13. Wilson is listed as mess treasurer in a list of collateral duties taped into Mustin's diary.
28 *Atlanta* Deck Log entry of February 11, 1942, NARA II.
29 *Atlanta* Deck Log entry of February 12, 1942, NARA II
30 *Atlanta* Deck Log entry of February 13, 1942, NARA II. Should there be a failure of the gyro compass, the ship would depend on the magnetic compass and since the magnetic north pole does not coincide with the Earth's true north pole, a difference in headings is expected.
31 *Atlanta* Deck Log entry of February 14, 1942, NARA II; Smith is listed in the collateral duties sheet inserted by Mustin in his diary.
32 Shaw, 84.
33 *Atlanta* Deck Log entry of February 17, 1942, NARA II.

34 McKinney, 22.
35 *Atlanta* Deck Log entries of February 18–20, 1942, NARA II; McKinney, 22.
36 *Atlanta* Deck Log entry of February 21, 1942, NARA II.
37 McKinney, 22–23.
38 Shaw, 84.
39 *Atlanta* Deck Log entries of February 22–23, 1942, NARA II.
40 Shaw, 85, has a complete transcript of the letter.
41 *Atlanta* Deck Log entries of February 26–27, 1942, NARA II.
42 *Atlanta* Deck Log entries of February 28–March 1, 1942, NARA II; Graff oral history.
43 *Atlanta* Deck Log entries of March 2–3, 1942, NARA II; Samuel Eliot Morison, *Two Ocean War: A Short History of the United States Navy in the Second War* (Boston, MA: Little Brown and Co. 1963), 88–101.
44 *Atlanta* Deck Log entries of March 4–6, 1942, NARA II.
45 Corboy, 5; "Diary of Unknown Sailor." Douglas St. Denis papers. A review of postwar U-boat loss lists show no sinking at this location at this time.
46 USS *Atlanta* Alumni Assn. Survey—M. D. Payton.
47 *Atlanta* Deck Log entries of March 6–7, 1942, NARA II; McKinney, 23.
48 Graff oral history, 15.
49 Shaw, 81–82.
50 "Diary of Unknown Sailor."
51 *Atlanta* Deck Log entries of March 13–14, 1942, NARA II.
52 "Diary of Unknown Sailor."
53 Shaw, 82–83; McKinney, 23.
54 "Diary of Unknown Sailor;" *Atlanta* Deck Log entry of March 18, 1942, NARA II.
55 *Atlanta* Deck Log entries of March 24–26, 1942, NARA II; Graff oral history.
56 *Atlanta* Deck Log entries of March 28–30, April 1-3, 1942, NARA II.

Chapter 3

1 *Atlanta* Deck Log entry of April 3, 1942, NARA II; Corboy, 5.
2 *Atlanta* Deck Log entries of April 3, 5–6, 1942, NARA II; USS *Atlanta* Alumni Assn. Survey—Charles Dodd.
3 Corboy, 5.
4 McKinney, 24. For the anchor pool there were 60 entrants each picking a minute of the hour and paying in a dollar. McKinney received 50 dollars as the organizer of the pool kept 10 dollars.
5 *Atlanta* Deck Log entry of April 8, 1942, NARA II; Jenkins letter to Commander in Chief Pacific Fleet dated April 15, 1942, *Atlanta* file Naval Warfare Branch, NHHC.
6 Timothy Boutoures, *You're in the Navy Now*, unpublished memoir, Navy Dept. Library. Bob Graff visited Balboa previously during his orientation cruise in *New York* and found that the local prostitutes ran from the officer candidates who went

ashore wearing dixie cup caps with a blue ring on the top edge. Apparently, the shore patrol had warned the locals that individuals wearing these caps had syphilis. Graff oral history.

7 McKinney, 24; Harry R. Mead, *Memoirs of a Pearl Harbor Survivor* (Washington, DC: Naval Historical Foundation, 2004), 12.

8 Richard D. Hepburn, *A History of American Dry-Docks: A Key Ingredient to a Maritime Power* (Arlington, VA: Noesis, Inc. 2003), 95. The Navy subsequently built bigger graving docks.

9 McKinney, 24; Enlisted Sailor Diary, 5; *Atlanta* Deck Log entries of March 11–12, 1942, NARA II.

10 *Atlanta* Deck Log entry April 12, 1942, NARA II.

11 Ibid; McKinney, 25; Graff oral history; *Atlanta* Deck Log entry of April 17, 1942, NARA II.

12 Graff oral history, 14; Unknown Sailor's diary, 5.

13 Ibid; *Atlanta* Deck Log entry of April 25, 1942, NARA II,

14 Corboy, 6.

15 Graff oral history, 18.

16 Unknown Sailor's diary, 5.

17 DANFS USS *California*, www.history.navy.mil. *California* lost 98 sailors during the attack.

18 Hawaii would revert to +10 hour in 1947. The quirk in the time zone at the time means that most annual Pearl Harbor commemorations remembering the moment when the first Japanese bombs fell are 30 minutes out of sync.

Chapter 4

1 John Prados, *Combined Fleet Decoded: The Secret History of American Intelligence and the Japanese Navy in World War II* (Annapolis, MD: Naval Institute Press, 1995), 302.

2 Diary of Lloyd Mustin. Founded in 1929, the company still makes high-pressure rotary pumps for the U.S. Navy. See: northern-pump.com.

3 Ibid.

4 Shaw, 87.

5 Wilbur D. Jones, Jr., Carroll Robbins Jones, *Hawaii Goes to War: The Aftermath of Pearl Harbor* (Shippensburg, PA: White Mane Press, 2001), 83–86; McKinney, 27.

6 McKinney, 27; The deck log recorded a summary court-martial for a sailor accused of covering up a venereal disease. He was acquitted.

7 Shaw, 87–88.

8 *Atlanta* Deck Log entry of May 6, 1942, NARA II.

9 *Atlanta* Deck Log entry of May 7, 1942, NARA II.

10 Ed Corboy, "The Log of the '*Mighty A*'", 6–6a; Unknown Sailor Diary, St. Denis papers.

11 Corboy, 6a; *Atlanta* Deck Log entry of May 10, 1942, NARA II.

12 Corboy, 6a.
13 Mustin Diary.
14 McKinney, 25–26.
15 Ibid.
16 Graff oral history; *Atlanta* Deck Log entry of May 14, 1942, NARA II.
17 Mustin Diary; Corboy, 6a.
18 *Atlanta* Deck Log entry of May 19, 1942, NARA II.
19 Corboy, 6a.
20 Corboy, 6a; Unknown Sailor Diary, St. Denis papers.
21 Mustin Diary.
22 Ibid.
23 Corboy, 7.
24 Mustin oral history; *Atlanta* Deck Log entry of May 26, 1942. NARA II recorded E. L. Graden as the recovered Sailor.
25 Unknown Sailor's diary, St. Denis papers.
26 Ibid. 98–102.
27 Morison, *The Two-Ocean War,* 145.

Chapter 5

1 Thomas B. Buell, *The Quiet Warrior: A Biography of Admiral Raymond A. Spruance* (Boston, MA: Little Brown and Company, 1974), 127–28.
2 McKinney, 26.
3 Mustin Diary.
4 Jonathan Parshall, "Admiral Nimitz's Planning for the Battle of Midway and Point Luck," presentation to Western Naval History Association Conference on February 19, 2022.
5 War Diary USS *Atlanta,* NARA II; Graff oral history, 21.
6 Shaw, 90.
7 Ibid. The transmissions were at 275 kilocycles.
8 War Diary USS *Atlanta,* NARA II.
9 Mustin oral history; *Atlanta* Deck Log entries of May 29–31, 1942, NARA II.
10 Mustin oral history.
11 War Diary USS *Atlanta,* NARA II.
12 Gordon Prange, *Miracle at Midway* (New York: Penguin, 1983), 163–64.
13 Prange, 174–76.
14 War Diary USS *Atlanta,* NARA II.
15 Prange, 244–45, 256–57.
16 Prange, 261–64, 270.
17 Corboy, 7.
18 Shaw, 90; Corboy, 7.
19 Corboy, 7.

20 Prange, 279–80; Corboy, 7.
21 Buell, 136–37.
22 Shaw, 91.
23 Prange, 286.
24 Shaw, 91
25 Prange, 286–87.
26 Prange, 290–91.
27 Corboy, 7a.
28 Buell, 139–40.
29 Mustin Diary.
30 Samuel Eliot Morison, *History of the United States Naval Operations in World War II: Coral Sea, Midway and Submarine Actions, May 1942—August 1942 (Volume IV)* (Boston, MA: Little, Brown, and Company, 1949), 144–45.
31 Prange, 325.
32 War Diary USS *Atlanta* NARA II; *Atlanta* Deck Log entry of June 6, 1942, NARA II; Morison IV, 148.
33 Buell, 142–43.
34 Prange, 339.
35 Shaw, 92.
36 Morison IV, 152; Prange, 357.
37 Morison IV, 149: Prange, 432–34; McKinney, 27.
38 Morison IV, 155–56.
39 Ibid.
40 *Atlanta* Deck Log entry of June 8, 1942, NARA II.
41 War Diary USS *Atlanta*, NARA II; *Atlanta* Deck Log entry of June 9, 1942, NARA II; Mustin oral history.
42 War Diary USS *Atlanta*, NARA II; Mustin oral history.

Chapter 6

1 Shaw, 92–93.
2 War Diary USS *Atlanta*, NARA II.
3 McKinney, 28.
4 Clifford Dunaway, Sr. oral history, *Atlanta* History Center Veterans History Project Oral History, YouTube posted January 27, 2016.
5 *Atlanta* Deck Log entries of June 16–17, 1942, NARA II. No mention of when such a party was held is made in the records.
6 *Atlanta* Deck Log entry of June 25–26, 1942, NARA II.
7 *Chock "A" Block*, August 6, 1942, 4–5.
8 *Mighty A*, 7a.
9 Corboy, 7a.

10 Mustin oral history.
11 *Atlanta* Deck Log entry of July 1, 1942, NARA II.
12 McKinney, 28.
13 *Atlanta* Deck Log entry of July 2, 1942, NARA II.
14 *Atlanta* Deck Log entries of July 4–5, 1942, NARA II.
15 *Atlanta* Deck Log entry of July 8, 1942, NARA II.
16 *Atlanta* Deck Log entry of July 10, 1942, NARA II.
17 *Atlanta* Deck Log entries of July 11–12, 1942, NARA II.
18 DANFS Vol. V. 108–09.

Chapter 7

1 Hixon, Carl K., *Guadalcanal: An American Story* (Annapolis, MD: Naval Institute Press, 1999), 7–8; Samuel Eliot Morison, *History of United States Naval Operations in World War II: The Struggle for Guadalcanal (Volume V)* (Boston, MA: Little, Brown and Co. 1949), 13–14.
2 *Atlanta* Deck Log entry of July 15, 1942, NARA II.
3 *Chock "A" Block*, Vol. I, Edition, 1.
4 Shaw, 96–97.
5 Graff oral history, 22.
6 *Atlanta* Deck Log entry of July 24, 1942, NARA II.
7 McKinney, 28.
8 *Atlanta* Deck Log entry of July 26, 1942, NARA II.
9 Shaw, 98.
10 *Chock "A" Block*, Vol. I, Edition, 1.

Chapter 8

1 Corboy, 8.
2 *Atlanta* Deck Log entry of August 7, 1942, NARA II; War Diary USS *Atlanta*, NARA II.
3 James D. Hornfischer, *Neptune's Inferno: The U.S. Navy at Guadalcanal* (New York, NY: Bantam Books, 2011), 40.
4 Shaw, 97; Enlisted man diary, 10.
5 Eric Hammel, *Guadalcanal: The Carrier Battles* (New York: Crown, 1987), 17–18. Hammel states 27 Bettys were in the first wave: Richard B. Frank, *Guadalcanal: The Definitive Account of the Landmark Battle* (New York, NY: Penguin Books, 1991), 68–69.
6 *Atlanta* Deck Log entry of August 8, 1942, NARA II; War Diary USS *Atlanta*, NARA II.
7 Hixon, 44.

8 DANFS, Vol. III, 505. Morison, *Vol. V*, 15–16. Hammel claims 18 Bettys and at least two Zeros were felled. Frank, 79–80, concurs.
9 War Diary USS *Atlanta*, NARA II.
10 Frank, 93.
11 Morison, *Vol. V*, 39; Hornfischer, 61; Frank, 108.
12 Morison, *Vol. V*, 46, 48, 54–55, 57–58.
13 Mustin Diary; Morison, *Vol V*, 61. The Navy did not announce the ship losses until 12 October when it triumphed at the battle of Cape Esperance.
14 McKinney, 29.
15 Mustin Diary; War Diary USS *Atlanta*, NARA II; *Atlanta* Deck Log entry of August 10–12, 1942, NARA II.
16 Unknown Sailor diary.
17 War Diary USS *Atlanta*, NARA II; *Atlanta* Deck Log entries of August 13–18, 1942, NARA II.
18 Hornfischer, 107: Frank, 126–27; Jeffrey R. Cox, *Blazing Star, Setting Sun: The Guadalcanal—Solomons Campaign, November 1942–March 1943* (New York, NY: Osprey, 2020), 147–48; Morison, *Guadalcanal*, 70–71.
19 Frank, 164.
20 War Diary USS *Atlanta*, NARA II; *Atlanta* Deck Log entry of August 23, 1942, NARA II; Frank 165.
21 Morison, *Two-Ocean War*, 179.
22 Frank, 176, War Diary USS *Atlanta*, NARA II.
23 Frank, 177.
24 Morison, *Two-Ocean War*, 180.
25 War Diary USS *Atlanta*; Frank, 179. There is a 30-minute time difference between contemporary narratives and *Atlanta*'s War Diary records the setting of Zone 11 ½ time on August 10 at 0623. Morison used 11 ½ time zone in his writings.
26 Frank, 180.
27 Cox, 175.
28 Mustin oral history; Corboy, 8a.
29 Mustin oral history; Frank, 182–83.
30 War Diary USS *Atlanta*; Corboy, 8a; Frank. 183.
31 Frank, 188.
32 Frank, 186; War Diary USS *Atlanta*.
33 War Diary USS *Atlanta*.
34 War Diary USS *Atlanta*.
35 Frank, 189–90.
36 Frank, 191.
37 Frank, 194.
38 USS *Atlanta* After Action Report dated Aug 27, 1942, Box 819, Record Group 38, NARA II.

Chapter 9

1 *Atlanta* Deck Log entry of August 29, 1942, NARA II.
2 War Diary USS *Atlanta*, NARA II.
3 Morison, *Two-Ocean War,* 182; Hornfischer, 121; War Diary USS *Atlanta*, Mustin oral history.
4 Frank, 204; Hornfischer, 121; Mustin oral history.
5 Hornfischer, 122.
6 War Diary USS *Atlanta*, NARA II.
7 *Atlanta* Deck Log entry of September 6, 1942, NARA II.
8 McKinney, 33.
9 Mustin Diary; Colleran and Underwood entries, 1943 *Lucky Bag.*
10 Ibid.
11 *Atlanta* Deck Log entry of September 20, 1942, NARA II.
12 DANFS *O'Brien.*
13 DANFS *North Carolina*; Frank, 249.
14 *Atlanta* Deck Log entries of September 21–23, 1942, NARA II.
15 Paul Stillwell, *Battleship Commander: The Life of Vice Admiral Willis A. Lee, Jr.* (Annapolis, MD: Naval Institute Press, 2021), 35.
16 Mustin oral history.
17 DANFS entries for *Vestal* and *Barnett*, NHHC; Mustin Diary.
18 McKinney, 28.
19 *Atlanta* Deck Log entry of September 26, 1942, NARA II.
20 *Atlanta* Deck Log entries of September 27–29, 1942, NARA II.
21 *Atlanta* Deck Log entries of October 3, 9, 1942, NARA II.

Chapter 10

1 War Diary USS *Atlanta*, NARA II; *Atlanta* Deck Log entries of October 9–10, 1942, NARA II; Hornfischer. 157; Mustin's diary.
2 Frank, 295.
3 Frank, 299.
4 Hornfischer, 169–72; Frank, 295, 300–01.
5 Frank, 305–06.
6 Cox, 288–89.
7 Frank, 308.
8 Frank 309.
9 Mustin Diary; Cox, 291–92.
10 Hornfischer, 193–95; Cox, 294–96; Frank, 316.
11 Cox, 298.
12 War Diary USS *Atlanta*, NARA II; Mustin Diary.

13 Mustin Diary.
14 Cox, 302–03.
15 War Diary USS *Atlanta*, NARA II.
16 Mustin Diary; Frank, 326.
17 War Diary USS *Atlanta*, NARA II; *Atlanta* Deck Log entry of October 18, 1942, NARA II.
18 Frank, 327; Cox 306.
19 160937 October 42 CINCPAC to COMINCH and 160245 October 42 COMINCH to CINCPAC located in Nimitz Gray Book Vol. II, NHHC.
20 Mustin Diary; Frank, 351.
21 Hornfischer, 219.
22 *Atlanta* Deck Log entry of October 19, 1942, NARA II.
23 War Diary, USS *Atlanta*, NARA II.
24 Cox, 314; DANFS *San Francisco*.
25 War Diary, USS *Atlanta,* NARA II; Mustin Diary.
26 War Diary USS *Atlanta*, NARA II.
27 *Atlanta* Deck Log entry of October 24, 1942, NARA II.
28 Cox, 321–24.
29 Frank, 365.
30 War Diary USS *Atlanta*, NARA II.
31 War Diary USS *Atlanta*, NARA II; Mustin Diary.
32 Mustin Diary.
33 Ibid.
34 Morison, *Two-Ocean War*, 193–94.
35 Cox, 326.
36 Frank, 374–75.
37 Hornfischer, 223–24.
38 Cox, 335.
39 Prados, 385.
40 Frank, 382–83.
41 Frank, 386; Cox, 348–49.
42 Hornfischer, 229–30.
43 Frank, 391–92; Morison, *Two-Ocean War*, 196.
44 Frank, 392–94.
45 Cox, 382; Frank, 396; Hornfischer, 232.
46 Frank, 397.
47 Frank 399; Cox 388–89.

Chapter 11

1 Cox, 393–95.
2 Mustin Diary; Mustin oral history; *Atlanta* Deck Log entry of October 27, 1942, NARA II.

3 War Diary USS *Atlanta*, NARA II; *Atlanta* Deck Log entry of October 28, 1942, NARA II. Mustin likely got to meet Scott during Mustin's 1200–1600 watch on the bridge.
4 DANFS, USS *Jacob Jones.* The second *Jacob Jones,* commissioned after World War I would suffer a similar fate on February 28, 1942, sunk by torpedoes fired from *U-578.*
5 Scott biography, Navy Dept. Library.
6 Hornfischer, 238; DANFS *San Francisco*; Callaghan biography, Navy Dept. Library.
7 Mustin oral history.
8 War Diary USS *Atlanta*, NARA II.
9 Mustin oral history.
10 War Diary USS *Atlanta*, NARA II; *Atlanta* Deck Log entry of October 29, 1942, NARA II.
11 Hornfischer, 243.
12 War Diary USS *Atlanta*, NARA II; *Atlanta* Deck Log entry of October 30, 1942, NARA II; Mustin oral history; Colonel Charles H. Nees, USMC, Biography courtesy Marine Corps History Division, Quantico, VA. Nees had already earned a Silver Star for his role in turning back a Japanese thrust to seize Henderson Field on the night of September 13.
13 Mustin oral history.
14 Mustin oral history; Hornfischer 243.
15 *Atlanta* Deck Log entry of October 30, 1942, NARA II.
16 Mustin oral history.
17 War Diary USS *Atlanta*, NARA II: *Atlanta* Deck Log entry of October 31, 1942, NARA II.
18 Hornfischer, 244.
19 Hornfischer, 244–45.
20 Hornfischer, 245–46.
21 Hornfischer, 247–48.
22 Colin G. Jameson, *The Battle of Guadalcanal, 11–15 November 1942* (Washington, DC: Office of Naval Intelligence, 1944), 4–5; Frank 431.
23 Mustin oral history.
24 Jameson, 9–10; Mustin oral history; Graff oral history, 24.
25 Mustin oral history. Mustin would reflect that wartime doctrine entrusted junior officers to make such decisions, even on a ship that carried an admiral.
26 Morison, *Vol. V,* 229; Mustin oral history; Graff oral history, 24.
27 Jameson, 8–9; Morison, *Vol. V*, 229.
28 Mustin oral history.
29 Jameson, 12; Mustin oral history.
30 Hornfischer, 251.
31 Mustin oral history; Jameson, 12.
32 Mustin oral history; Hornfischer, 252; Frank, 431.
33 Mustin oral history.
34 Frank, 431; Mustin oral history.

Chapter 12

1 Jameson 16; Morison, *Vol. V*, 236; Mustin oral history.
2 Hornfischer, 260; Jameson, 11.
3 Morison, *Vol. V*, 239.
4 Frank, 437.
5 McKinney, 43. Of note, RBAs came into use in the Navy in the 1930s and during the Cold War this equipment became better known as the Oxygen Breathing Apparatus (OBA). Oxygen is generated using an inserted canister containing potassium superoxide which reacts initially to a sodium chloride candle and continues to react as the wearer exhales carbon dioxide. The Navy began replacing this system in 2001.
6 Mustin oral history; Jameson, 12.
7 Hornfischer, 268; McKinney 43; Dunaway oral history. Mustin posted the *Atlanta* Watchbill in his diary.
8 Mustin oral history.
9 Hornfischer, 269; Frank, 438.
10 Hornfischer, 269. Mustin oral history.
11 Mustin oral history.
12 Hornfischer, 273–74.
13 Mustin oral history; Dunaway oral history.
14 Hornfischer, 275–76.
15 Hornfischer, 274.
16 Reed oral history.
17 McKinney, 46.
18 Mustin oral history.
19 McKinney, 46.
20 McKinney, 47; USS *Atlanta* Alumni Survey—Oscar Ekberg.
21 Shaw, 106–07; McKinney, 47.
22 McKinney, 47; USS *Atlanta* Alumni Survey—Wayne Langton; USS *Atlanta* Alumni Survey—Kenyth Brown.
23 Mustin oral history.
24 Mustin oral history.
25 USS *Atlanta* Alumni Survey—Leighton Spadone; McKinney, 47. One of Spadone's gun crew, William Simpson, would suffer a spinal wound that would require permanent care.
26 McKinney, 44; USS *Atlanta* Alumni Survey—Bill McKinney.
27 Stephen M. Younger, *Silver State Dreadnought: The Remarkable Story of Battleship Nevada* (Annapolis, MD: Naval Institute Press, 2018), 3; Mustin oral history.
28 McKinney, 44.
29 Mustin oral history.

30 Shaw, 107.
31 Mustin oral history.
32 Reed oral history; Graff Oral History, 26; Commander Emory letter dated November 18, 1942, to Commanding Officer, USS *Atlanta*, NARA II.
33 Shaw, 108.
34 Shaw, 107; Dunaway oral history; Reed oral history; Alvin Otto oral history conducted by Christopher Willoughby on July 21, 2005, Library of Congress Veterans History Project.
35 Morison, 252.
36 Mustin oral history.
37 Shaw, 108.
38 Shaw, 109.
39 Mustin oral history.
40 USS *Atlanta* Alumni Survey—Bill McKinney; Commander Emory letter dated November 18, 1942, to Commanding Officer, USS *Atlanta*, NARA II.
41 USS *Atlanta* Alumni Survey—Bill McKinney.
42 McKinney, 46; Harvey memoir, 18.
43 Harvey memoir, 18.
44 Jameson, 30–32; Morison, *Vol. V,* 254.
45 Harvey memoir, 18; Hornfischer, 322; Morison, 254–55; Mustin Oral History.
46 Hornfischer, 324–25.
47 Harvey memoir, 18.
48 Mustin oral history.
49 Commander Emory report dated November 18, 1942, NARA II; Harvey memoir, 18.
50 McKinney, 49.
51 Frank, 459; Mustin oral history.
52 Harvey memoir, 19.
53 McKinney, 49.
54 Ibid, 51
55 Mustin oral history; Morison, *Vol. V*, 255–56.
56 McKinney, 50.
57 Frank, 456–59.
58 Ibid, 457.
59 McKinney, 51.
60 Ibid, 52.
61 Hornfischer, 332; Mustin oral history; Harvey memoir, 19.
62 USS *Atlanta* Assn. Alumni Survey—Wayne Langton.
63 McKinney, 52.
64 Harvey memoir, 20.

Chapter 13

1 McKinney, 55; Reed oral history.
2 Mustin oral history.
3 McKinney, 55.
4 Corboy, 13a.
5 Mustin oral history.
6 Hornfischer, 337–38; Frank, 464.
7 Mustin oral history; Harvey memoir, 20.
8 McKinney, 55; Reed oral history.
9 Frank, 464–65.
10 McKinney, 56.
11 Frank, 465–66.
12 Frank, 468–69.
13 Mustin oral history.
14 Mustin oral history; Frank 472.
15 Frank, 475.
16 Frank, 476–77.
17 Frank, 484–85. *Gwin* recovered *Benham*'s crew and scuttled her with gunfire the following evening.
18 Frank, 479–80.
19 Frank, 481.
20 Frank, 483.
21 McKinney, 57.
22 McKinney, 57; Frank, 487–88; Mustin oral history.
23 Mustin oral history.
24 Frank, 488. In addition to the soldiers, some 1,500 bags of rice (a four-day supply) and 260 boxes of shells were landed.
25 Shaw, 111.
26 Shaw, 111.
27 McKinney, 58.
28 McKinney, 58.
29 Harvey memoir; Mustin oral history; McKinney, 59.
30 McKinney, 60.
31 McKinney, 60; Harvey memoir.
32 Harvey memoir; McKinney, 63.
33 McKinney, 63. McKinney identified the Sailor as M. J. "Frenchy" Lamothe.
34 DANFS USS *President Monroe*. The Navy would acquire the ship on July 18, 1943, and place her in commission a month later with a Navy crew.
35 McKinney, 64; Shaw, 112; Reed oral history.
36 McKinney, 64.

37 McKinney, 65.
38 McKinney, 66–67; Corboy, preface to "*The Log of the 'Mighty A.'*" Before the series ran, Captain Jenkins and other officers from the *Atlanta* wardroom had the opportunity to review the content and Captain Jenkins provided introductory commentary in the initial installment in the series that ran on March 20, 1943. The series would be syndicated and featured in papers across the country.
39 Shaw, 112.
40 Shaw, 112–13.
41 Commander Emory action report to Commanding Officer USS *Atlanta*, dated November 18, 1942, NARA II; Shaw, 114–17.

Chapter 14

1 Preston Grady, "Cruiser Bond Drive Headed by Conner—10,000-Tonner Now on Ways Will Bear Name *Atlanta* Vinson Predicts," *The Atlanta Journal,* January 13, 1943, 1, 6; Western Union telegram from John Conner to Al Sharp dated January 12, 1943, FDR Library; Conner letter to Grace Tully, Secretary to the President dated January 12, 1942, FDR Library.
2 DANFS, *Atlanta* (IV).
3 Al Sharp, "Cruiser Campaign Sweeps Over Top," *Atlanta Constitution,* CL-104 Files Ship History Section, NHHC.
4 Al Sharp. "*Atlanta* Cruiser Fund Passes $63,000,000 On Last Day of Drive," *Atlanta Constitution,* CL-104 Files Ship History Section, NHHC.
5 Garver Files, Operational Archives, NHHC. Of note, after the war he would be assigned to Parris Island and would drown on September 10, 1946, apparently having fallen from his sailboat off Parris Island.
6 Erdman File, Operational Archives, NHHC. Erdman would retire as a captain on January 1, 1974.
7 Jessie Fant Evans, "Impressive Ceremony Is Held As Cruiser *Atlanta* Joins Fleet," CL-104 Files Ship History Section, NHHC.
8 Don K. Love letter to W. Graham Claytor dated April 18, 1979; W. Graham Claytor letter to Ivan Allen Jr. dated June 11, 1979, CL-104 Files Ship History Section, NHHC.
9 DANFS *Atlanta* V.
10 DANFS, *Birmingham*; Shaw, 198–200.
11 Shaw, 126.
12 Shaw, 255; Shaw Files, Operational Archives, NHHC. His December 6, 1988, *New York Times* obituary that noted his having passed two days earlier, labeled him "Admiral and Historian."
13 Nickelson Files, Operational Archives, NHHC.

14 Colleran Files, Operational Archives, NHHC.
15 Jenkins Files, Operational Archives, NHHC; "Lowry Report Absolves 'Green Bowlers' of All Charges; 18 Admirals listed. *Armed Force* (October 1, 1947).
16 Emory Files, Operational Archives, NHHC; DANFS, *Miami.*
17 Sears Files, Operational Archives NHHC; Malcolm W. Cagle, Frank A. Manson, *The Sea War in Korea* (Naval Institute Press, 1957), 41–43, 93.
18 Wulff Files, Operational Archives, NHHC.
19 Mustin Files, Operational Archives, NHHC; John Fass Morton, *Mustin: A Naval Family of the 20th Century* (Annapolis, MD: Naval Institute Press, 2003), 291.
20 Morton, 291–93.
21 McEntee Files, Operational Archives, NHHC.
22 Luddy Files, Operational Archives, NHHC. His online obituary stated he passed away on May 10, 2011, at Mystic CT, having been married to Rita Fitzgerald who passed in 2007.
23 USS *Atlanta* Association file for Hall; *Oakland II* (CL-95), DANFS, NHHC.
24 Trever Jensen, "Edward D. Corboy 1919–2007," The *Chicago Tribune,* May 1, 2007; Graff oral history.
25 USS *Atlanta* Association files for Smith, Dodd, and Ashlock.
26 USS *Atlanta* Association files for Spadone and McKinney.
27 *National Geographic: Lost Fleet of Guadalcanal,* National Geographic Society, 1993. In 1995, Kevin Denley and Terrence Tysall first dove on the ship and returned in subsequent years. In 2011, divers from Global Underwater Explorers made six dives to record footage for the documentary *Return to the USS Atlanta: Defender of Guadalcanal.*
28 Graff interview with author.

Bibliography

Archives

Naval History and Heritage Command (NHHC), Washington DC

Ship History Files
Officer Biography Files

National Archives, New York (NARA NY)

RG 80 General Records of the Department of the Navy

National Archives, College Park (NARA II)

USS *Atlanta* Deck Logs
USS *Atlanta* War Diaries

Seeley G. Mudd Manuscript Library, Princeton University, Princeton, NJ

Harold G. Bowen Papers

Douglas St. Denis papers

Diary of Unknown Sailor
H. Charles Dahn, "The Death of the Mighty A," paper written December 28, 1958
USS *Atlanta* Reunion Survey

Books

Bowen, Harold G. *Ships, Machinery, and Mossbacks: The Autobiography of a Naval Engineer.* Princeton: Princeton University Press, 1954.

Buell, Thomas B. *The Quiet Warrior: A Biography of Admiral Raymond A. Spruance.* Boston, MA: Little Brown and Company, 1974.

Cagle, Malcolm W. and Frank A. Manson. *The Sea War in Korea.* Annapolis, MD: Naval Institute Press, 1957.

Corboy, Ed. *The Log of the Mighty A*. USS *Atlanta* Reunion Scrapbook, 1985.

Cox, Jeffrey R. *Blazing Star, Setting Sun: The Guadalcanal–Solomons Campaign, November 1942–March 1943*. New York, NY: Osprey, 2020.

Frank, Richard B. *Guadalcanal: The Definitive Account of the Landmark Battle*. New York, NY: Penguin Books, 1991.

Friedman, Norman. *U.S. Cruisers: An Illustrated Design History*. Annapolis, MD: Naval Institute Press, 1984.

Hammel, Eric. *Guadalcanal: The Carrier Battles*. New York, NY: Crown, 1987.

Hepburn, Richard D. *A History of American Dry-Docks: A Key Ingredient to a Maritime Power*. Arlington, VA: Noesis, Inc., 2003.

Hixon, Carl K. *Guadalcanal: An American Story*. Annapolis, MD: Naval Institute Press, 1999.

Hone, Trent. *Learning War: The Evolution of Fighting Doctrine in the U.S. Navy, 1898–1945*. Annapolis, MD: Naval Institute Press, 2018.

Hornfischer, James D. *Neptune's Inferno: The U.S. Navy at Guadalcanal*. New York, NY: Bantam Books, 2011.

Jameson, Colin G. *The Battle of Guadalcanal, 11–15 November 1942*. Washington, DC: Office of Naval Intelligence, 1944.

Jones, Jr., Wilbur D. and Carroll Robbins Jones. *Hawaii Goes to War: The Aftermath of Pearl Harbor*. Shippensburg, PA: White Mane Press, 2001.

Luraghi, Raimondo. *A History of the Confederate Navy*. Annapolis, MD: Naval Institute Press, 1996.

McKinney, William B. *Join the Navy and See The World*. Los Angeles, CA: Military Literary Guild, 1990.

Morison, Samuel Eliot. *History of the United States Naval Operations in World War II: Coral Sea, Midway and Submarine Actions, May 1942–August 1942 (Volume IV)*. Boston, MA: Little, Brown, and Company, 1949.

Morison, Samuel Eliot. *History of the United States Naval Operations in World War II: The Struggle for Guadalcanal (Volume V)*. Boston, MA: Little, Brown and Company, 1949.

Morison, Samuel Eliot. *Two Ocean War: A Short History of the United States Navy in the Second World War*. Boston MA: Little, Brown and Company, 1963.

Morton, John Fass. *Mustin: A Naval Family of the 20th Century*. Annapolis, MD: Naval Institute Press, 2003.

Prados, John. *Combined Fleet Decoded: The Secret History of American Intelligence and the Japanese Navy in World War II*. Annapolis, MD: Naval Institute Press, 1995.

Prange, Gordon. *Miracle at Midway*. New York, NY: Penguin, 1983.

Scharf, J. Thomas. *History of the Confederate States Navy*. New York: Gramercy Books, 1996.

Shaw, Elizabeth R. P. *Beside Me Still: A Memoir of Love and Loss in World War II*. Annapolis, MD: Naval Institute Press, 2002.

Stillwell, Paul. *Battleship Commander: The Life of Vice Admiral Willis A. Lee Jr.* Annapolis, MD: Naval Institute Press, 2021.

Sulzberger, C. L. *World War II*. Boston MA: Houghton Mifflin Company, 1966.

Winkler, David F. *Ready Then, Ready Now, Ready Always: More Than a Century of Service by Citizen Sailors*. Washington, DC: Navy Reserve Centennial Committee, 2014.

Wright, Captain Richard L. *To Provide and Maintain a Navy, 1775–1945*. Arlington, VA: Strategic Insight, 2018.

Younger, Stephen M. *Silver State Dreadnought: The Remarkable Story of Battleship Nevada*. Annapolis, MD: Naval Institute Press, 2018.

Dissertations

Reynolds, Charles V. Jr. "America and a Two Ocean Navy, 1933–1941." PhD. diss., Boston, MA: Boston College, 1978.

Oral Histories and Memoirs

Boutoures, Timothy. *You're in the Navy Now*. Unpublished memoir, Navy Dept. Library.

Dunaway, Clifford, Sr. Oral History, *Atlanta* History Center Veterans History Project Oral History, 2016.

Graff, Robert D. Interview by author, dated December 27, 2003.

Mead, Harry R. *Memoirs of a Pearl Harbor Survivor*. Washington, DC: Naval Historical Foundation, 2004.

Mustin, Lloyd M. Interview by John Mason. U.S. Naval Institute Oral History Program, 2003.

Reed, Franklyn LeRoy. Interview by Michael Willie of December 20, 2004, Library of Congress Veterans History Project.

Styles, Ralph. Interview by Earl Fowler, Naval Historical Foundation Oral History Program, 2001.

Diaries

Diary of Lloyd Mustin.

Articles

Peterson, Maw C. "Naval Courts Martial." *Indiana Law Journal,* Vol. 20, Issue 2 (Winter 1945), 168–69.

Index